QUIT FEEDING
THE
MONSTERS

CREATING A POSITIVE CULTURE

QUIT FEEDING THE
MONSTERS

CREATING A POSITIVE CULTURE

J. KEVIN COBB

Two Harbors Press
212 3rd Avenue North, Suite 290
Minneapolis, MN 55401
612.455.2293
www.TwoHarborsPress.com

ISBN-13: 978-1-936401-45-1
LCCN: 2011922272

Distributed by Itasca Books

Cover Design and Typeset by Wendy Baker

Printed in the United States of America

Foreword

I have known Kevin for almost fourteen years now. We have worked together on hundreds of implementations. Since I've written a few things myself (one of them has sold over 1 million copies), I tried to help Kevin with the question that anyone who's ever thought about writing a book has asked themselves: "Who am I to be writing a book? I don't have formal academic qualifications, research studies, or data…"

What I told him was this: You have spoken to more employees in small groups than almost anyone else in the world. What you have learned talking and listening to over 250,000 people *is worth reading!* Don't worry about research. It rarely explains what really works when the rubber meets the road anyway. But experience—real life experience—usually does. And it provides people with more confidence because they know that what you're saying really worked for someone, somewhere, sometime. So don't worry about credentials or research, just tell it like it is and let your readers form their own conclusions.

It is also worth mentioning that Kevin is a gifted storyteller with an incredible memory for detail. So you get to feel what it was really like to be out there on the front lines of the battlefield where the war against the "monsters" is being fought. Sometimes the monsters win and sometimes the people slaying the monsters win. That's just reality in the modern corporate world. And Kevin doesn't pull any punches.

So, take what you will from this book. It's a great read, but it also has a powerful message.

Larry Wilson
Author of *SafeStart*

Acknowledgments

Writing a book was a tremendous undertaking and could not have been done without the generous help and support of many.

That effort has produced a great sense of personal reward; however, it would be wrong to infer that somehow *Quit Feeding the Monsters* is mine alone.

The efforts of many are in your hands and I want them to know publicly what their contributions to this book and my life have meant.

Above all, I thank God for putting the project on my heart and for putting up with me when I questioned His judgment on why this was needed and for asking constantly if He was sure He had the right man for the job.

Special thanks to Larry Wilson. If it were not for him, this book would not exist. His encouragement, coaching, feedback, and support made sure this part of my dream came true. Larry is a gifted monster slayer and I'm proud to call him a friend. Larry and his entire team at Electrolab Training Systems are a big part of all the success I have had.

Special thanks to Dave Threlfall for all his chats throughout the years. Dave, you have served as one of life's wonderful teachers;

thank you for answering all my crazy questions.

Sometimes I'm amazed at how many people contribute to a project like this and how they contribute. Mike Stacy did not directly have a hand in this book, but his fingerprints are all over it. Mike, thanks for the inspiration and leading the way a leader should. It was Mike who first showed me the way of the Monster Slayer.

Thanks to Mike Reeder for his initial edit of *Monsters*. Mike, you set out to make my manuscript better while promising to protect the integrity of my vision. You did both far beyond my expectations.

Thanks to Rachel Fichter, my editor at Two Harbors Press, who surely at times must have wondered if English was a second language for me. Your attention to detail has served the *Monsters* well.

Thanks to Cathy O'Keefe for her business partnership throughout the years. Cathy's tremendous efforts have afforded me the time to discover and question the way of the Monster Slayer. Cathy, I know it's not always easy. You've been a tremendous teammate.

Ruth Steeves and Jack Jackson, thank you both for your encouragement and support. Your validation of the monsters concept was a tipping point for me. I can never thank you enough.

Thanks to Bill Blalock. It was a conversation with Bill in Baton Rouge several years ago that was the actual genesis of the *Monsters*. At the time I never dreamed where that chat would take me.

Thanks to Miss Kristi who puts up with a crazy husband and allows me to chase my dreams at a tremendous personal cost. I will never fully understand the sacrifices you've made or why you chose to do so, but without them, my life would be so much less.

ACKNOWLEDGMENTS

You are a wellspring of selfless support. I don't know how you do it.

Finally, I want to thank my dad. I cannot explain what he means to me as a father, leader, and teacher. He is ultimately the very soul of this book. To go through this process with him has been one of the highlights of my life. Thank you, Dad, I love you.

Dedication

This book is dedicated to the men and women who do the work day in and out; my dad, my first leader and a gifted manager; my sister Danna, who signed on as the first Monster Slayer; and my children, who are the greatest blessing of my life. Thanks, Alyssa, Rose, Riley, and John Eli.

Table of Contents

Quit Feeding the Monsters...

Once upon a time, in the land of business, a lad set out to free the vast kingdom from the grip of its ugly monsters.

The monsters were having their way, creating havoc and threatening the very existence of the kingdom itself. They had become quite skilled in their acts of terror; the fear of them and their unchecked power left many with little hope that someone would rise to conquer the beasts.

As the battles waged on, the Monster Slayer grew ever disheartened. While he was skilled in finding the monsters and helping others destroy them, he soon realized that many of the very people who wanted to kill the monsters were actually nurturing the beasts—sometimes with no idea they were doing so.

If he were truly able to win this epic battle, he would have to take on an unexpected foe: he had to stop those feeding the monsters.

Starving out the monsters seemed to be the thing to do, but those protecting the beasts would not let them die so easily. In fact, many fought to keep that which they hated alive.

While it was clear what had to be done, the true challenge lay in how to go about it. You see, the monsters had been allowed to grow for so long that they had taken on many forms.

While the monsters had many faces, they were easy enough to recognize. Some saw the monsters as mean, others saw them as political; fact is, the monsters could be both. They could also be lazy, manipulative, rude, unprofessional, irrational, prone to fits, divisive, loud, obnoxious, unreliable, backstabbing, quick to blame others, self-centered, lying, deceitful, boastful, and cruel as well as crude, hurtful, bullying, sly, argumentative, and angry. More insidiously, they could be charming, friendly, highly successful, and so much more.

Taking on such beasts was not going to be easy. The Monster Slayer would have to develop and perfect certain skills in order to defeat such a multi-faceted foe. He would also have to count on others in his quest to conquer the beasts. He had to somehow rely on those feeding the monsters.

*　　　*　　　*

Okay, it reads a bit like a kid's bedtime story, but eliminating the so-called monsters that are destroying your business is hardly kids' play.

Years ago I started my career helping companies that wanted to improve their safety performance and, in turn, their employees' lives. I found it to be rewarding work that I was proud to do.

The company I worked for was offering real solutions that had tremendous value—as my boss at the time put it, "A noble business." Noble indeed. We were partnering with companies not only to ensure a safer workplace but to improve safety off the job as well. We were working to make a real difference in protecting employees and their families. When an employee suffers a serious injury, his or her family's emotional and financial health is also a casualty.

Serious accidents could, in fact, wreak havoc on families

for generations. Think about a young child who loses a mother or father. What kind of adult will that child become without that loving, guiding hand, and what kind of parent will he or she become later?

To say the least, it was a job I took very seriously. At times I found professional joy that I could have never previously imagined. We all have a need to give, and I found mine. It was real, meaningful, and was clearly helping families.

I eventually realized, however, that many of the people hiring us were, in fact, the biggest challenge to our and their own success.

Don't get me wrong. They were trying to help keep employees safe. But their actions and, many times, their lack of action were proving tough hurdles to overcome, and the effects were being felt far beyond safety performance.

Not long into my career, I realized that the basic building blocks of safety also applied to productivity and quality. So if we could improve just a few things overall, performance should sky-rocket.

While in most cases employees' behavior needed to change, that wasn't likely to happen as long as managers and supervisors continued to mismanage their most valuable asset: the employees themselves.

This mismanagement took many forms. The most troubling was managers who enabled employees who frankly held back the organization.

Quit Feeding the Monsters was born out of frustration. Senior managers were constantly telling me that their mid-level managers and employees were the problem. Supervisors, for example, did not know how to manage people.

They had, I was repeatedly told, little skill when it came to conflict resolution, teambuilding, coaching, mentoring, and so on. The obvious question was, "What have you done to develop those skills?" The answer more often than not was, "Well...nothing."

Often, supervisors and mid-level managers were good employees who were rewarded with a position for their efforts and years of service. Of course I'm not opposed to taking care of those who take care of us, but being a good employee for ten to twenty years does not make a good leader. It's a great start, but leadership requires much more than technical know-how.

It seemed to me that senior management was complaining about a problem only they had the authority and resources to solve. And thus the idea of "Feeding the Monster" was born. If things were really going to improve, managers had to quit reserving the right to complain about the very problems they were creating and, in many cases, enabling.

This book did not start out as a book. It was bits and pieces of wisdom and, in some cases, lack of wisdom that I began to pick up from good and bad managers alike. I also realized that I had already learned much of the valuable information I was passing along simply by watching my father deal with people throughout his nearly fifty-year management career.

When the idea for a book came about, the question that soon followed was, "What makes you qualified?" The answer was, "Well, nothing really." Okay, so you're going to write a book on a subject you're really not qualified to write about? Sort of, I guess, if by qualified you mean long years in management. But I also knew from first-hand experience that many simple truths were being ignored or overlooked by far too many managers, MBAs, and old-school leaders alike.

More importantly, I've spent a lot of time with employees who have suffered under bad management and some who were inspired to new heights by those leaders who somehow just "got it."

My job as a safety consultant has taken me to some of the toughest environments imaginable—off-shore drilling, utility companies, general manufacturing, forest products, and automotive production just to name a few.

My career has taken me to hundreds of different companies and well over 500 individual worksites. Getting a chance to see the industrial workplace from that perspective gave me opportunities to see patterns that are just not visible at one site or company level.

Many see the failure of their leadership and suspect other companies are run much better. Some are. More importantly, the patterns that determine success or failure are consistent.

There were consistent behaviors practiced by leaders at companies that did well and a different set of behaviors practiced just as consistently by those companies that failed. It was those patterns that grabbed my attention. These were not random behaviors happening by chance. After looking at hundreds of organizations and many more worksites, the road to success was as clear as the path to failure.

It's that unique perspective that few ever have the benefit of enjoying over the course of their careers that allowed me to first determine these patterns and, more importantly, verify them over time and in hundreds of organizations. That perspective is ultimately my qualification for writing this book.

Interestingly enough, these patterns also crossed cultural boundaries. My work has taken me to every state in the United States, nearly every province in Canada, the Artic circle, and all over Mexico.

Regardless of the type of work or location, the management mistakes and victories all seemed to remain the same. As an outsider, it was clear that I had a vantage point that allowed me to see things others either could not see or had dismissed as being "just the way things are around here."

My job as a consultant was simply pointing out good and bad management behavior alike, most of which the employees were pointing out to me.

I've felt many times that companies pay consultants to tell

them things their own employees would tell them for free. I am thankful that many companies to this day have not discovered their own staff's cache of ideas and unique perspectives. The consulting business is alive and well because of it.

This book is really a by-product of what happens when companies foster a climate that cuts off communication. Employees simply don't feel they are allowed to tell their managers what "the real story is around here."

This book is what employees would tell managers if employees thought they would listen. My guess is that if an employee actually told his or her boss stuff like this, it would only happen as he was turning in his two weeks notice, perhaps hoping that a moment of honesty might prompt management to let him get on with his new life right away.

The techniques of good leaders vary and are in a constant state of change to meet the ever-evolving demands of the workplace. It makes sense. Good leaders are always changing and reinventing themselves and testing new keys to success.

This book was not primarily written for senior executives. I do not doubt, however, that they would also find quite a bit of valuable information in these pages. This book is for those front-line managers and supervisors; those who lead workers day in and day out.

It's also worth mentioning that this book is not the end-all, be-all on leadership. It was never intended to be a repository on the subject. It is, rather, a collection of stories and experiences that should serve as learning tools for you as they were for me. The purpose of this book is simply to establish some baseline principles we can all benefit from when it comes to leadership.

This book is for supervisors and managers who want to lead their teams successfully. It's also for those who just want to learn how to be better leaders.

Formal leadership training in most organizations does not

exist. This book is for those who are not happy with the status quo or for those who are not sure who they are as managers and are seeking their way—a way that will catapult them to new heights as leaders, all while helping those they manage, or will one day manage, to achieve their personal and professional goals.

A word of caution to those of you about to enter the realm of the Monster Slayer...this is not a business for the faint-hearted.

You'll be challenged at every turn. Killing organizational monsters is fraught with danger and political pitfalls that 007 would find difficult to negotiate. What I'm trying to say is, don't get so caught up in the task of monster slaying that you forget that your workplace exists to make money and provide service.

The following pages are going to suggest some radical changes. They are going to call into question some long accepted business practices. You are about to become a change agent. You will know what it feels like to be the person rocking the boat. You will step on some toes and you will likely not be the most popular person for a while. But the war is long and there are many battles before a final victory is declared.

Take time to learn from your successes and failures. Build consensus where it makes sense. Hold the line as best you can when the situation calls for it.

Remember, you're out to slay monsters that may have been developing for decades. A quick, decisive win is the stuff of kids' storybooks.

Be patient and stay the course. You won't be much of a Monster Slayer if you get fired two weeks into the battle.

Slaying monsters is noble work, but you still have to feed the family, pay the bills, and prepare for those golden years when you can sit back and reminisce about your days as a Monster Slayer.

GETTING WHAT YOU WANT

"Men occasionally stumble over the truth, but most of them pick themselves up and hurry off as if nothing had happened."
—Winston Churchill

One night, at age four-and-a-half, my daughter Riley snuggled up in my arms and announced that I was the greatest daddy ever and that she loved me. My heart melted. She followed it up with, "Even when you're mean to me!"

Being a dad is the best job I've ever had, but it's demanding and it has challenged me like nothing else in life. It has produced moments of great triumph and joy and, at times, what seemed like heartbreaking losses.

Sometimes it was tea parties, attending recitals, camping or ski trips, or just painting nails. Other times it was making sure my kids understood that "no" meant "no."

I don't think anyone is a natural born parent. I believe there is a great deal of skill that's learned on the job, and for the most part, you've got to go out and learn that skill through trial and error.

Now, there's plenty of help out there. Everyone has an opinion on how to raise a kid, and it seems there are more experts in the field than there are kids. But one lesson stuck very early with me and it came from noted psychologist Dr. James Dobson. The question Dr. Dobson presented to his audience was simple and profound. It would eventually change me as a father, manager, and a man.

The question was, "Dad, do you thank your children when they do what they are supposed to do?" My honest response in my heart was, "Of course not!" When my kids do what they're supposed to do, it's because that's what's expected; no thank you needed, end of subject.

To me, that was their job, and the expectation on my part was that things should be done because I said so. If I needed to say it more than once, I wasn't happy.

I was their boss, they worked for me, and their room and board was their pay.

Let me make sure his point is not lost on you as it initially was on me. Should parents thank their children for making their beds, taking out the trash, cleaning their rooms, and knocking out homework without being asked twenty times?

Every parent has been there, having to tell his or her child what seems like hundreds of times to get the simplest of chores done. But occasionally children do what they are supposed to do on their own. Again, the question was, in those cases, did you thank them for it? For me, the honest answer was no. Hey, nobody was coming up to me and saying thanks for the food and the house payment each month, let alone vacations, investments, and all the other things I do for my family.

I never stopped to think what I was really saying to my kids or what I was teaching them. If you want a behavior repeated, it makes sense to positively reinforce it with a simple thank you. Dare I even praise them?

You see, when my children failed to do as they were told, I was ready to pounce on them. But when they demonstrated the behavior I wanted, it often went unrecognized. As their mother once said, everything can't be World War III with the kids. You've got to pick and choose your battles.

The more I thought about positive reinforcement, the more convicted I felt as their father.

It just made sense that reinforcing the good things and positively correcting things that did not go well was going to make my job as their dad more effective and much easier for all of us. I had, on at least this front, failed them as a father. That had to change. I was sure of that, but I had yet to fully appreciate just how profound that little question posed earlier really was.

Think about it. How would you like to be treated? I've told that story several times to leadership groups and have been surprised that some in the audience chose to take from it that I see employees no differently than children.

Actually, it would be more accurate to say I see some similarities between lousy management and less-than-stellar parenting. After all, both parenting and management are really just about leadership.

I have struggled most in my life over decisions regarding my children, making sure I was focused on the big picture, not getting lost in squabbles of the moment, all the while trying to lay out a plan that would one day lead to their success.

It's all leadership, being a dad, a mom, a good boss, a go-to employee, or a trusted friend.

The realization that the feedback we get may not match the reality of how we actually perform is not exactly new, but it is astonishing how widespread it is in our society.

Think about all the meetings you've had with your boss. Were the majority of your meetings centered on thanking you and celebrating the good job you do day in and day out, or was it more along the lines of being told that you're not quite measuring up? By the way, are you a screw-up most of the time? My guess is you're just like the rest of us. For the most part, you do a good job. So the question is how has the feedback been along the way?

And it's just not a workplace issue. Our society does a really lousy job with this positive reinforcement/positive correction approach.

Have the police ever pulled you over and thanked you for driving safely? That is, after all, what you're doing most of the time. Okay, they can't constantly pull us over and thank us (it would get a little annoying). But how about just once!

If you're married or have been married, how much positive reinforcement have you received as a husband or wife, and what about positive correction between couples? The latter idea is laughable to most spouses.

Leadership is sometimes about correcting behavior, but some act as if that's all there is to it. Is that true?

If you thought back over your entire career, how much of the time, in terms of percentages, would you say you preformed well? Eighty percent of the time? Ninety percent of the time?

Look, we're not perfect. No one scores 100 in life, but for the most part, most of us perform well. My guess is that 80 to 90 percent of the time, if not more, most of us perform our jobs well. So the question is, "How's the feedback been?" Has it reflected the reality of how well you've performed, or has it been pretty much one-sided?

Most of us get a "talking-to" when things go wrong, but where's the talking-to when things go right? I'm not suggesting we should go around blowing sunshine at people, but it would be nice to get some recognition for the things we do well. Furthermore, when we do need some correction, where is it written that it must include belittling people?

Real, meaningful, positive reinforcement for a job well done goes a long way when it comes to correction time, especially if we could find a way to do that in a respectful manner too.

Years ago, a boss of mine attended a training session I was

leading. He was to lead a group right behind me. He showed up early and sat in on my session. I showed up the next morning and found a simple note expressing his pleasure with the job I had done. It was only five words. I still have it with me today. That single little note meant so much to me that I never threw it away.

As our careers went on, there were times that I did not do so well and correction was needed. He was equally gifted in this task as well. By the way, when I needed to be corrected, it was much easier to take because of his penchant for positive reinforcement. In short, I found him to be fair all of the time and gave him the benefit of the doubt even when I disagreed with him.

My other choice would be to dismiss his criticisms, but that would force me to discount all the times he handed out a "job well done." I could not in good faith dismiss one without dismissing the other. I decided to keep them both.

So, if we are performing 90 percent of our tasks well, we should be receiving at least that much positive feedback, if not more. It's been said you need to hand out ten "at-a-boys" for every "ah-shit."

While I've never studied sociology and don't have anything resembling a scientific answer as to why most of us are this way, I have come across several theories.

"Well, that's the way I was treated, and now it's my time with the hammer." I actually had an oil field manager tell me that once. He told me he had never received any positive feedback about his work and he turned out just fine. Actually, he was angry about nearly everything and quite argumentative. But okay, he turned out fine.

He really struggled with the concept of positive feedback and asked, "Since when do roughnecks have feelings?" My guess is, they always have and always will.

I asked him how he felt when he was treated roughly, and, again, he insisted that he turned out all right. But the question was how did it feel? He continued to insist that it was his turn to bully people and that he had earned the right to do so. He was not going to give it up, especially now that it was finally his turn to swing the hammer.

So, if anything, this was sort of payback time. Trouble is he was paying back those who had nothing to do with the way he was treated and helping create the next generation of lousy bosses.

He eventually capitulated, but in his mind—at least initially—hammering his guys was the right thing to do because he had been hammered.

Another manager balked at the idea of positive reinforcement/correction, saying to do so would be a disaster for the company.

He admittedly had bullied employees for years and felt if he changed now they would not respect him or his decisions. I told him, "They don't respect you now; fear and respect are not the same thing." He insisted that kicking employees was the way he got what he wanted. If he didn't get what he wanted, he simply wasn't kicking hard enough.

He considered a positive reinforcement/correction approach "soft," even referring to it as a bit "Oprah-ish." I don't think that was supposed to be a compliment. Though, I don't ever recall telling him to hug his employees, grab a cup of coffee, or join a book-of-the-month club.

If he really wanted some respect, he was going to have to start showing some, and given the amount of disrespect he had been dishing out over the years, he would likely have to wait a while to see a return on his respect investment.

While it's certainly not the point of this book, respect could serve as its general theme. Seems to me there is far too little of it left in our society. We disrespect authority, institutions, and each other. We have become the culture of me and now. That includes "leaders" who believe it is the employees' job to serve them versus the other way around.

A boss I truly admired as a young man told me and my fellow employees that it was his job to get rid of the organizational roadblocks that were holding us back. It was his job to make sure we had a clear path to success. In short, it was his job to serve us.

When leaders actually serve, something amazing happens. People flourish and develop new talents, allowing everyone to benefit. Leaders who want their "needs" met would be wise to meet the needs of those they lead.

For many, this positive reinforcement/correction approach is a wonderful way to meet some of those needs and inject some respect back into the conversation.

Little Riley had it right; in many ways I am a wonderful father—in her estimation, "the greatest Daddy ever!" I could, however, improve my approach to discipline. Admittedly, I have a tendency to overreact at times. So in her eyes, I just needed to improve on the whole losing-my-head thing.

Hmm…out of the mouths of babes! My four-and-a-half year-old and a noted psychologist had figured out something that had been completely lost on me. Thank you, My Little Love! GREAT JOB!

PEOPLE DEVELOPMENT BUSINESS

*"If your actions inspire others to dream more, learn more, do more
and become more, you are a leader."*
—John Quincy Adams

Working with a client at a sawmill years ago, I came across a truly inspirational management team. The meeting to roll out a new safety process was scheduled to last three entire days.

I had asked the mill manager to assemble his lead team.
Now, such requests are usually met with the "rolls down hill" effect. Department heads avoid the meeting by "volun-telling" their direct-reports to take their places in the meeting, because running the department just can't tolerate that kind of time drain.

After all, the department heads are far too important for such a time commitment. Besides, the supervisor and, indeed, the organization would likely benefit more from a supervisor attending because he is actually closer to the troops. Quite a clever way to avoid a meeting, right?

Much to my surprise, as we worked our way around the conference table that morning making introductions, I found my audience not to be typical mid-level managers at all. I not only had the mill manager himself, but managers of production, safety, maintenance, quality, and even the company's comptroller, along with the rest of the entire executive staff.

So we were already off to a great start. After years of consulting and hundreds of these sessions, I had never seen such a gathering of executive-level leadership for a three-day process meeting.

As I began our meeting with some simple housekeeping issues, I was not only impressed by, but sensitive to the fact that we had the entire organizational brain trust held hostage for the next seventy-two hours. So I announced that I had no problem with people leaving their cell phones on in case of an emergency. I only asked that they set them to stun.

I was a bit taken aback by the mill manager who glared at me and barked, "Everyone turn off your phones! There is nothing more important in these next three days than what we're here to talk about."

While I appreciated the importance he had just assigned our gathering, I recall thinking, "Lighten up! It's not exactly brain surgery on the agenda." But he was right, even if I felt a little like a child being scolded by a parent. What's the point of being in a meeting if you are not focused, learning, and contributing?

Look, we've all seen or heard those in the back of the meeting who are busy answering e-mails or playing games on their phones. First of all, it's rude, not to mention distracting to everyone else. And it sends a loud and clear message that the perpetrator thinks he and what he does is more important than any meeting. When that is indeed the case, we should excuse that person from the meeting.

This was not the last time during those three days that I would be impressed with this particular manager. I wouldn't give him much for style points, but he always seemed right-on with his assessment of his team, where they were, and when to get them back on task.

As often happens in meetings, people tend to get off track. Soon the discussion has little, if anything, to do with the reason why there was a meeting in the first place. Many folks handle such misdirected conversation with the "parking lot" approach. In other words, the misguided issue is set aside for detailed discussion later.

This manager, as he often did, had a bit of a different approach. When his managers got off task he would bark, "What business are we in?" (I know I keep using the word "bark," but it's the only one that accurately describes what I witnessed.) In response to that, his team would chant (it seemed in unison), "We are in the people development business." His response would then be, "Get back to the people's business."

During one of our breaks I asked him to explain the "people development business."

He told me that many people mistakenly thought that they were operating a sawmill. Looking around and seeing lumber, sawdust, debarking machines, and saws, it was easy to see how someone could get confused. I didn't have the courage to tell him he was nuts.

He continued, "Our job as a leadership group is to develop people. Find them and match their talents with our needs. We find out where they want to be in five, ten, or fifteen years. Our job then is to make sure they get the skills training and professional development they need to reach those goals. So when the opportunity arises, we already have the right people in place.

"So if you are in maintenance and you desire to be the maintenance manager, it's our job to make sure we get you the leadership tools to go along with your technical competence."

He was also quick to point out that some people are right where they want to be and that's okay. As he put it, "We need great maintenance people too." The whole focus was on providing employees with the tools to reach their goals.

"As a matter of fact," he said, "I'm currently helping to develop several people who want my job. So, we are in the people development business. By the way," he added, "the by-product of that business is cutting the wood."

Wait a minute. An executive manager and an entire organization that really cares about the long-term goals of all its employees and was willing to back it with a real plan versus the old "work hard and you'll get there." Now there's a new concept for you.

I never really warmed up to this guy's personality, but as the years have gone by, I've thought about how I'd work for him anytime, any place. Wouldn't you?

See, I perceived our meeting to be about rolling out a new process, yet he saw it as one more step in his people development strategy.

You can buy good leadership or you can develop it and keep it all to yourself. The latter tends to be more cost effective and therefore more profitable.

Too many times I've seen good managers and employees leave for better long-term opportunities. It seems no one was at all interested in what these people's long-term goals were. When they left, all the talent that many times was developed on that very job was allowed to just slip right out the door.

Grooming employees takes time and a tremendous amount of money. Having people walk because there was no plan to meet their needs is equivalent to opening the flood gates of your organization. The outpouring of intellectual capital is staggering.

The fact is you cannot hire off the street the intellectual skills you helped developed. With a new hire you have to start all over. Think about it. A new employee is not even sure where the restroom is.

People are constantly leaving organizations. However,

research indicates most people are leaving their bosses, not the company itself. Many times they may even be leaving the monsters that are being fed by poor leadership. It could very well be more cost effective to have the right leadership in place versus trying to constantly bring new people up to speed.

In today's performance-demanding world, our goals have become too short term. What are we doing this quarter or this year? Getting a long-term focus is a wonderful way to make sure your workplace is not a breeding ground for monsters.

Unless employees are leaving because of retirement or termination, you should be fighting to keep them and the intellectual capital your organization paid to develop in them.

And since eventually someone *will* leave, you must have a plan to replace everyone on your team, including yourself, just like our mill manager. For this very reason, many have turned to formalized mentoring (more on that later), but nothing beats strong leadership that actively cares about employees' long-term goals.

LET'S STEP OUT BACK

*"Our chief want is someone who will inspire us to be what
we know we could be."*
—Ralph Waldo Emerson

Good managers tend to make good decisions, and, in turn, their
employees tend to develop a great deal of confidence in their
bosses' ability to assess a situation and react appropriately, and at
times, good managers just flat out inspire.

Working in the oil and gas industry has always been a great
pleasure for me. The men and women (mostly men) in that industry
seem to have a rugged, independent, can-do spirit about them.

It reminds me of some of the Western movie heroes I grew up
watching, with a heap of foul language to boot. They tend to be the
most straightforward people you'll ever meet. Most of them have
no problem giving you their two cents' worth and will not trouble
themselves with cleaning it up for a sensitive set of ears. There's
something about that kind of candor in this day that is refreshing.

For what they may lack in genteel attributes they more than
make up for in honesty and hard work, and they work in some of
the harshest environments imaginable.

J. KEVIN COBB

Getting this tough-minded, independent crowd to move in an organizational sense takes a strong, decisive leader. But what if that leader could also inspire? Well then, you'd see some amazing things.

I met one such manager while consulting on a new safety process. Rodney was in his mid-fifties and a divisional manager for a global drilling company. He had all the traits of a typical man who had spent his life in the oil patch. He was brimming with confidence and a let's-get-it-done attitude that I'm certain some mistook as cockiness.

We had met through some mutual associates, and, after hammering me with tons of questions about the new process, he thought what I had to say merited his guys' attention. Rodney, however, is the type of manager who empowers his people to make decisions. In turn, he holds them accountable for those decisions.

"Give them the power and responsibility and you'll get their best effort" was his line of thinking.

So Rodney told me to meet him in Oklahoma City and he would allow me to speak to his seven zone safety managers and he would charge them with the final decision of whether or not they would move forward. He wanted their buy-in, and if I couldn't get it at that meeting, we were done.

While we'd had several discussions in the past, I really had no idea just how committed Rodney was to his job and his people. I was about to find out. As he introduced me, he reminded his safety managers that it was their job to protect his 400 brothers and sisters who were working under their leadership in their division of this global company.

Referring to his employees as his brothers and sisters told me a lot about Rodney's commitment to those who worked for him, but I was about to learn much more. Rodney reminded his team that if anything should go wrong on their watch, they would ultimately be held accountable to him.

He said, "I have knocked on a door to tell a wife and her children that her husband and their father would never come home again. If it happens on your watch, you will go with me and face that broken family. I will never do that alone again.

"After that, you will attend every meeting or anything else connected to an OSHA investigation." They investigate all workplace fatalities. "If legal action should follow, you will attend all meetings with lawyers and you will be there for every minute of the trial.

"Once all of that is done," he concluded, "your troubles are just beginning, because you and I are going out back."

Rodney handed the group he had just threatened over to me and left the room. Not knowing for sure how to take all of this, I had to ask, "Was that real?"

The response from one of his managers was, "The threatening to whip our ass part? Sure, but he hasn't actually had to do that in years." The rest of the room chuckled.

I'm not sure to this day if Rodney ever actually laid a hand on any of his guys, and that's not the point. Rodney believed in something and he made sure his people believed what he believed. More importantly, he required ownership and accountability of his vision of how his organization should conduct its business.

The managers I met with that day were wonderfully critical. I can't remember a time before or since when I was asked so many pointed questions. They were a strong group of men who took their business very seriously.

By the way, that seriousness showed Rodney's division led the entire organization in safety, productivity, and quality performance. There is no question in my mind that his leadership was a big key to their success.

Rodney later summed it all up this way: "In our business the competition is tight. We all offer the same service. We all use the same equipment and we buy it from the same vendors. In this kind of business, separating yourself too much on price point could kill you.

"In fact, the only thing we really have to offer that our competition doesn't is our people. So we make sure our people have the right tools, training, and leadership to out-perform anyone in the business. And we measure every bit of it. Why? Because when we bid a job we always come in high, and we don't have a shortage of business, because we are able to show our clients that we are their best bet for safety, quality, and productivity.

"It just makes sense to pay more for us. You'll make more money long term, pure and simple. It's a better value."

What a great way to position your organization in the marketplace. Of course we cost more. You get us! And we're the best any way you want to measure it.

Rodney told me, "Without our people, we are nothing. If my team fails our people, then we have failed completely."

Well, we got the project, and when I got out in the field to work with those who spent their days on the rigs, it was no surprise that I found fierce loyalty to Rodney and that his leadership fingerprints were all over those 400 people. They felt and talked openly about how it was their job to protect their brothers.

Each member of that crew, all the way down to the newest rig hand, had the authority to shut down the entire rig if they saw something that could get someone hurt.

Rodney told me, "Even if they're wrong, I'll back their decision publicly, and if they were wrong, we'll take that up privately one-on-one. These guys and everyone out there needs to know that I stand by them."

Okay, I want to make sure you got that. Everyone on that job had the authority and responsibility to make a decision to shut down the operation, a decision that could potentially cost tens if not hundreds of thousands of dollars.

Now lots of people say they have that kind of leadership, but their employees tend to roll their eyes when asked, "Could you really do that?"

I asked many of Rodney's guys if they really believed they had the authority to shut down a big-dollar job and still expect support from management. To a man they said, "Yes," and many added, "It better be real, or you have to take it up with Rodney."

Granted it's not a formal survey. But when that many people respond with that type of conviction, you have got to believe something is there.

Speaking at a leadership conference once, I related Rodney's story. During the question-and-answer session a lady in the audience commented that that oil field guy was barbaric.

I responded in a way I hope Rodney would appreciate. "Ma'am," I said, "at times the oil field is a barbaric place. But you still have to perform, and that requires strong, committed leadership, the kind that will inspire some pretty tough-minded individuals. Call it barbaric or committed. Either way, it works for Rodney and his brothers and sisters."

Leadership styles clearly differ from manager to manager and from industry to industry, but top-performing organizations have leaders who inspire their employees to perform above standard, and that is exactly what Rodney does.

Style isn't important here, inspiring employees is. If your efforts inspire people to push themselves in the right direction and beyond what they think they are capable of, then, like Rodney, you're leading well.

IS IT ALL ABOUT LEADERSHIP?

"Where there is no vision, the people perish."
—Proverbs 29:18

The way I see it, a large part of organizational success is tied to good leadership, so it should come as no surprise where I lay most of the blame when things go poorly. You have to take the good with the bad, right?

It starts with a vision that comes from leaders, followed by accountability. With no vision, there can be no accountability. With no vision and no accountability, there is no real leadership.

Just about every organizational failure is ultimately a failure of leadership. Leaders are hired to lead, and when the organization does not perform up to standard, managers are to blame, kind of harsh, huh? It may be harsh, but the truth is often harsh.

The only group that has the power to drive an organization is the leaders.

Employees fail for many reasons, most of which are easily attributed to their managers. The manager failed to lay out a vision, failed to communicate that vision and the employee's role in achieving it, failed to motivate, failed to hold anyone accountable, and so on and so on and so on.

Wait a minute, Kevin! Some employees are jerks who are unworkable! I'd have to agree with you on that one, and I have indeed met a few incorrigible employees in my travels. But it's still a failure of leadership not to move that person to a more productive attitude or role for the overall good of the organization, either through coaching, mentoring, or discipline.

Discipline, by the way, has severe limitations and should be a last-ditch effort. More to the point, it's the beginning of the termination process.

Once the positive things have been done and you still have a problem employee, then it's because you failed to terminate.

Often I hear, "I inherited the problem." Well, you still have to find a way to motivate. And if you can't do that, then you need to remove the problem from the organization.

Fact is, as you change jobs in your career, you will inherit all the problems and sins of the previous leadership group and you'll get no credit for their victories.

You will be judged initially as being either as sorry as the last group or nowhere near as good as the last group was. You're the leader now and you inherit all the bad stuff. It is, after all, your job now to deal with it.

Over the years, I have been fortunate enough to have many capable mentors—some of them have even become friends. One friend, Gary, makes this point wonderfully.

Gary had just taken over as a new plant manager. He called a meeting of his management team that was to begin promptly at 9 a.m.

At nine, Gary found himself alone in the meeting room. He had suspected this would happen. In his short time at the helm, he had already seen the drag-ourselves-to-a-meeting mentality.

Gary, true to his word, began the meeting right on time. That's right: Gary was addressing his staff as scheduled despite the fact he was the only one in the room.

About ten minutes into the meeting, the first manager showed up to find Gary talking to an empty room. He quietly and uncomfortably took a seat with Gary hardly acknowledging that he had walked in.

I would have loved to know what was going through that guy's mind as he walked in to see Gary talking to an empty room. In the next few minutes, others trickled in.

At the end of the meeting, Gary told his managers that each of the assignments that he had given out at the beginning of the meeting was due on his desk by the close of the business day, no exceptions. With that, he simply walked out of the room.

About a half hour later, he heard a sheepish knock at his door. They had selected their sacrificial lamb to ask, "Just what was it you wanted us to do again?"

Gary handed him the team's assignments and, amazingly, from that moment on, if someone was late for a meeting, they always had a representative there to make sure notes were taken and nothing was missed.

Gary has a PhD in Monster Slaying. And letting folks think you might be just a little bit crazy tends to keep them on their toes, too. Thanks for that lesson, Gary.

*   *   *

The sawmill leadership group that I mentioned earlier had its faults. We all do. One was the failure as a group to deal with a long-term employee they all considered a problem.

The safety process we were rolling out had several key components, including the idea of positively reinforcing desired behavior and positively correcting at-risk behavior.

The leadership was worried about one thirty-year employee who was highly knowledgeable regarding his job and the mill, reliable, never missed work, but was the most negative guy in the entire mill.

So their question for me was, "While we agree with your 'positive' approach, what would you do with this employee who hates everyone and everything?" Or as they actually put it, "He's going to (expletive) all over this positive approach stuff."

It seemed like a silly question to me at the time and does even now as I relate it to you. Ultimately, there is a correlation between the talent one brings to the organization and the amount of BS the organization should tolerate from that individual.

If this gentleman was indeed everything they claimed he was, then I guess I would just deal with him and ride it out to his approaching retirement. Either his value far outweighs the amount of stress he adds to the organization, or he's not that good and he's a bigger drain on everyone than he is worth.

I was in no position to answer their question, so I returned it to them with, "Why bring a problem (everyone agreed it was a problem) to me that you've had thirty years to fix?" It was one of the few times an indictment was met with laughter. Assess the problem from a risk/reward perspective and make a decision.

My guess is they had done that informally and that this man was more than worth the grief he was causing, which makes their question about how to handle his reaction a moot point. That said, I'll talk a bit more later on about how to manage difficult employees. In many cases, they are otherwise great employees. Often, all that is keeping them from being the complete employee you really want is leadership.

So, the so-called problem they were asking me about was simply a failure of leadership to assess the pros and cons and respond accordingly (coach, contain, or eliminate). Sometimes this failure can be directly connected to the problem. Other times it's more like a dotted line.

BAD EMPLOYEE...BAD

"Get your facts first, and then you can distort them as much as you please."
—Mark Twain

A union leader I was working with asked for my opinion about a recent accident one of his men had experienced. An employee had used a stepladder to perform a task for which he actually needed a six-foot ladder. So the employee got by, so to speak, by standing on the very top of the stepladder where a label clearly says, "Do not stand here."

The official accident investigation was an open and shut case. The employee used the wrong tool for the job. No discipline was handed down, but the blame was clearly put on the employee and written off as a poor choice on his part. Makes sense until you ask some questions.

The first one should be obvious: does the employee have access to the six-foot ladder? The answer was yes, but he had to go to the tool room to get it.

By the way, the tool room was about 100 yards away. Okay, so when he needs the ladder, it's 200 yards round trip. The rules also required that he return the ladder to the tool room after each use. So it's really 400 yards round trip each time he needs the ladder.

My next step was to ask how often during the day he would need this ladder to perform the task. I was told up to half a dozen times a day and that he was required to check the ladder in and out of the tool room each time.

Let's see, that's 400 yards per use, six times a day, for a daily total of 2,400 yards. Then, multiply 2,400 yards five times for each day of the week. My goodness! That's over 10,000 yards! This guy is covering more yards in a week than most NFL running backs accumulate in a career. No wonder the stepladder was tempting!

It seemed to me that had the employee been provided with a ladder in a handier location he would likely have used it. When asked why this was not done, I was told it would not be possible under current company policy.

Six-foot ladders and several other tools had been identified by the management team as "hot" items. "Hot" meaning they had a history of disappearing from the plant. In other words, they were being stolen.

The response to this thievery was to tighten control on certain tools. So each time one of the tools that management knew employees were stealing was needed, employees had to check it out and back into the tool room immediately after each use.

I asked if management had any idea who was behind the thefts and was told, "Sure, we have them on video tape." So I asked, "What was done with the evidence?" The answer was "locking up the tools." No one, he told me, had been fired for stealing.

When I asked him why the clearly guilty parties, caught stealing on tape, were not terminated and handed over to local prosecutors, the answer was astounding. "We can't fire them," he said. "Many of them are supervisors!"

The gentleman relating the story was making a point, of course, and I was hearing him loud and clear. Now back to our incident with the stepladder. What we have here is a failure of leadership, not an employee using an improper tool, as the actual root cause of the accident.

When we build systems that add needless work and frustration to employees' lives, we are creating monsters. Seems to me it would have been easier in the long run to get rid of the thieves. What was kept instead was a frustrating system designed by "leaders" who refused to lead.

In addition, they had unwittingly designed, thanks to their own lack of leadership, a system that got someone hurt. They had also created an environment that lacked accountability, which in turn created a breeding ground for even more monsters: if they catch you stealing and you're still not fired, what else could you get away with around here?

That very thought drains your good employees and feeds the monsters you should be eliminating. Good people don't steal, but each day they're forced to work with those who do without punishment slowly chips away at their morale. That only produces more monsters down the line.

My friend Gary puts it this way: "If I can't trust you with little things, what else can I not trust you with?" Not showing up on time for meetings and stealing from the organization should be telling you something. Are you listening? Do you hear the monsters setting up house?

LITTLE MONSTER FALLING OUT OF THE SKY

"It's easy to see, hard to forsee."
—Benjamin Franklin

Sometimes the monsters are obvious. When you have thieves, they need to be eliminated. Sometimes, though, monsters silently undermine the organization.

Once, when I was working with a forest products company, there seemed to be a general practice of not following some select rules. Now, it was hardly anarchy, but not following some established rules, policies, and procedures undermines the integrity of all the rules and policies.

I was asked if employees should follow one rule in particular that required them to wear hard hats at all times. Seems like an easy question on the surface.

Incidentally, I learned long ago that these seemingly innocent questions usually have a much deeper, darker purpose.

So, should they wear the hard hats? "Depends," I answered. I love to watch people roll their eyes at that response. I asked that they explain a little more.

It seems crewmen planting trees as part of the reforestation plan had decided they would not follow the rule. Seemed reasonable to me; they were just planting seedlings. After all, the only thing above their heads was the bright, beautiful British Columbia sky. Interior BC is a beautiful place!

The easy answer was that employees should follow all rules at all times, but it did seem to be a silly rule for these planters. I guess the only thing they had to worry about over head was meteorites, but come on! At that point, I doubt the hard hat would help much.

Now this is where it gets a little more complicated. The supervisor of the planting crew was tired of hearing his teams complain about the seemingly silly rule, so he decided to relax it a bit.

He decided, without trying to change the rule first, that the guys on his job would not have to comply. Easy, right? Problem solved and he likely figured no one would know what his guys were doing out in the forest anyway.

Problem was, word spread, and as it did, others in the company began to complain that they too should not be required to obey the rule. I mean, it did apply to everyone right?

That led to others deciding which rules were important and which ones were not. The only problem is that leadership should be responsible for making those kinds of decisions. Yet leadership only spoke up when the monster was out of control, and the response was hardly what I would call measured or appropriate.

Instead, leadership came up with a new rule that stated, "Follow all the previous rules." That, of course, meant all rules, including those that may not make sense.

Let's back this up a little. Twenty years ago, no one at this company wore hard hats, not even those who clearly needed them.

The company was worried about the Resistance to Change

Monster. Instead of meeting that straight on with measured and strong leadership, it took the path of least resistance, which was to make a blanket rule that applied to everyone, even if it did not fit.

That's a position I can *almost* talk myself into. The quickest way to change the climate in hopes of changing the culture is the no-exception rule. Note I said quickest, not most effective.

They clearly had needed a change and it had to stick. The problem came later, when employees began pointing out how silly the requirement was for those working with nothing but blue sky above.

Instead of reevaluating and backing off when appropriate, leadership stuck stubbornly to its guns and said just do it because it was a rule.

The wisdom of the hard-and-fast rule is questionable at best. But if that is the course you choose, then you have an obligation to revisit the issue and reevaluate once you have either achieved the overall behavioral change desired or someone points out an obvious deficiency, whichever comes first.

All policies and procedures need to be reviewed and updated from time to time. Once hard hat compliance was no longer a glaring issue, the leadership group should have modified the rule.

Because they did not, they created the perfect climate for those who were just waiting for the opportunity to point out that this leadership group had no idea what happened in the real world.

That, of course, put every decision leadership made in question. At least it did with the monsters. They were only too happy for the help in tearing down the company and eroding faith in its leadership.

We have a responsibility to constantly revisit our decisions and to evaluate what they mean in current context. Just because the decision was right or almost right once upon a time does not mean it's right for all time.

If there are rules you don't agree with because they are not necessary or don't make sense, you need to step up and initiate the process to get things changed. Failing to do so will only lead to bigger monsters later because you decided to feed them by doing nothing.

LEAD! THEY WILL FOLLOW

"If you command wisely, you'll be obeyed cheerfully."
—Thomas Fuller

Leaders, lead! Okay, nothing earth shattering there. But it is amazing how many leaders fail to understand that what they do can bring down or lift up an organization. Wherever you lead, trust me, they will be watching.

Because of that, when working with a new company, I feel it is always critical that I spend time with the leadership team.

I hold these meetings to make sure that all leaders understand my expectations and goals and what their part in the process will be. But it's also about building support and answering the age-old question, "What's in it for me"? If I could answer that question, implementations were much easier.

If I take the time to build a consensus and to answer the questions and concerns about what I'm doing, then the leadership group can prepare its team for what's coming.

With everyone informed, I have much more buy-in going in. That buy-in at launch is priceless. If you don't get enough of it, most improvement initiatives crash and burn shortly after taking off.

A brilliant sales consultant once shared this computer sales adage with me: "Long sales process, short install. Short sales process, long install."

The longer the sales process, project, or anything else, the more buy-in you need and naturally collect. The more buy-in along the way, the easier and more productive the road becomes for all.

People need to understand their roles, have their concerns addressed, and their needs met. You can only do that with well-thought-out leadership. The old, I'm-the-boss-get-it-done mentality may get it done, but there's little hope of sustaining anything because employees lack ownership. They need a shared vision.

All people are capable of some role as leaders. In that sense, I need leadership, ownership, and accountability from each and every team member.

Leaders can be found at all levels of an organization, and many of these leaders don't have titles. However, they, not the executive group, often decide what really does and does not get done.

Who are the leaders you work with? The men and women who get things done may not hold positions of authority, but, at least in part, they hold the key to your ultimate success. Identify them; pull them into your projects. Elicit their advice and ask for their talents. Some of the best ideas are had by leaders with no titles.

Take your time with any new initiative. Take time to develop a unified leadership group to drive the new effort. I've seen many companies rush implementation with little communication about what's being required and what the expectations are.

If your focus is on long-term, sustainable success, then don't rush the implementation. Otherwise, it will likely be another "flavor of the month."

BEAT THOSE GOOD HORSES

"People are definitely a company's greatest asset. A company is only as good as the people it keeps."
—Mary Kay Ash

Human beings by nature seem to take the path of least resistance. No one likes to deal with an unpleasant co-worker. Instead of dealing with the problem or taking on problem people, we typically take the easy way out and simply insist on more out of our top performers.

At every organization I have ever worked for, it was easy to identify who the "organizational quarterbacks" were. You know who they are. They are the men and woman who get things done.

They know what they're doing and they mostly get it done right the first time without a lot of direction and without whining like a child who has been handed an extra chore.

Identifying these people is so easy, just about anyone can do it. Their easy-going natures and get-it-done attitudes stand out like rays of hope. Unfortunately, it's not long before everyone who needs something knows who to turn to. In short, it's a dog pile.

We abuse their time, talent, easy-going attitudes, and burn

them out long before they should.

Meanwhile, the monster who receives a "good morning" and responds with a growled "What's good about it?" is left to his or her own devices, and little extra is ever asked or expected of him or her.

I'm not talking about personality differences here. People are different and that diversity of outlooks and talent makes it all work. I'm talking about those who use a horrible attitude as a shield against work.

I remember one quarterback in particular. John was not only technically competent. He was a joy to work with. He was always calm and positive, even in the high-paced and often chaotic environment of live television news.

I was in my early twenties and new to the business, and John was a life preserver of control and competence in a swirling sea of confusion.

You could always count on John to keep his head on straight and our team focused. But it came with a heavy cost.

John once told me he wanted to change his name so he would no longer have to answer to, "Hey John, could you help me?" Certainly, John and all of us have a responsibility to learn to say no. Great leaders understand their limitations and know that the more they stretch their time and talents, the less effective they are. But great leadership also needs to identify those who can't seem to say no and, in some cases, say no for them.

John really could have used some help from leadership, but the company demanded more from John than any of us, and not just a little more.

John, while pleasant, was a stressed out bundle of nerves. In large part because of the stress, he had become a chain smoker. But for a myriad of reasons, he had decided it was time to quit smoking.

After several failed attempts at quitting, John enlisted the help of a hypnotist. The way it all worked, he explained, was that a hypnotic suggestion would be planted in his subconscious that cigarettes and snakes were the same thing.

The idea, he told me, was to find something that terrified you then equate that, through hypnotic suggestion, with the idea of smoking. You crave a smoke and you think snakes, which in turn produces fear and a shot of adrenaline, which helps you through the craving. And snakes were helping John beat his addiction. He feared them that much.

One day, I stepped outside of the building and was surprised to see John smoking. He had an ash well over an inch long on the end of his cigarette, which meant he was hitting it pretty hard.

John looked at me, smiled, and said, "I'm not afraid of snakes anymore." The stress of the job was so powerful and so overwhelming, it had finally overpowered the hypnotic suggestion! On the upside, John would never fear snakes again.

John never did develop the heart to say no and management continued to dump tons of work on its star employee. By the way, all the while, less talented, far less pleasant, and out right lazy employees in similar positions were allowed to coast through their days with little asked of them.

John eventually had too much and quit in hopes of finding a less stressful life. We, of course, were left with an angry, complaining, and far less talented group of people to take his place.

I've seen it over and over. Instead of dealing with problem employees, good employees tend to be ridden into the ground. Many times they feel trapped. To say no is a risky proposition (or so they rationalize), so they just continue to burn themselves out, delivering a level of performance that should never have been demanded of them in the first place.

It makes much more sense to raise the level of everyone's performance and to ease the pressure on our quarterbacks. Instead, they leave, and many times we simply go out and hope to find another great horse to whip.

Another friend of mine is a very talented graphic artist. She happened to work for a university. She loved her work and was very good at it. The problem was that everyone wanted to benefit from her talent, including departments she didn't belong to.

Her desire to "pitch-in" led to an onslaught of new projects, demands, and deadlines. Hey, when there's a new workhorse, word spreads fast.

It's very difficult to say no to a dean of a college even when you don't actually report to him or her. To make matters worse, the dean she reported to felt free to volunteer her time and talent to other departments for projects that had nothing to do with what she was actually held accountable for.

She had a penchant for making him look good when he "loaned" her out to others, but it soon backfired on both of them.

The work quickly piled up and her days and weeks got longer and longer. In predictable fashion, all of her projects began to suffer.

. The whole thing came to a head when some important and very public deadlines were missed. She was in trouble with her boss who (of course) created most of the problem, and she was soon looking for another job. Not to further her career, but to escape her boss.

Of course, in her department there were others who could have handled some of the additional work, but because she was so pleasant, not to mention good at what she did, everyone took the path of least resistance and overwhelmed her to the point that she wanted out.

All the while, those who contributed much less stayed on as a financial and morale drags on the organization, fat, happy monsters.

FIRE 'EM!

"I mean, there's no arguing. There is no anything. There is no beating around the bush 'You're fired' is a very strong term."
—Donald Trump

Once while working with a client, he pointed out that over the course of the day I had suggested twenty-three times that someone be fired.

It was hypothetical of course, but he wanted to know how I could believe so much in coaching, mentoring, and leadership while at the same time, over the course of a day full of meetings, the suggestion to fire someone came out of my mouth twenty-three times. I was actually amazed myself that I had said it that many times, and what was he doing counting, anyway?

Okay, it had become sort of a joke that day. They would ask a question and I would flippantly suggest they just fire the problem and move on.

I never found out why he bothered to count the number of "fire 'ems" coming out of my mouth, but I'm glad he did because it caused me to stop and ask myself why I had suggested so many times, even jokingly, that the best course to take was to let these people go.

First of all, let me say that when I've personally had to fire someone, I have always viewed it as a failure of my own leadership. Either I failed to coach, motivate, lead, inspire, correct, or discipline. If I did all I could do and the employee was unreachable, then I failed to hire the right person.

It is always a failure of leadership unless, of course, you inherited the problem. Then it is a failure of the previous leadership. But it soon becomes yours if you fail to remove the person and the risk he or she represents to the organization. Like I said, it is always a failure of leadership.

I have been fired and I have fired. Neither was enjoyable or something I want to experience again, though I suspect, one way or another, I will.

To let someone go is a difficult decision, but leadership is about difficult decisions.

As an airline pilot once said to me, "I could have you flying this very plane in five minutes. You don't pay pilots to fly planes. Computers do that. You pay pilots to get them back on the ground safely when things go wrong."

The same holds true for leaders. Managers have plenty of systems and tools and people to run things. Good leaders are the ones that get it all back on track when the organization or individuals are not performing well.

Good leaders are, of course, careful about whom they hire. Beyond that, they tend to take it upon themselves to serve the needs of those they work for.

Leadership is about taking care of your people. They, in turn, should take care of the managers' and organization's goals. After all, our goals should be mutually aligned.

Leaders, first of all, need to hire the proper people. Okay, I'm stating the obvious here, but they are the only ones who have the authority to do the hiring. I won't detail how to hire people. There is plenty of material already available on the subject.

I will, however, say this. Make sure you talk to the prospect several times. Have others whose judgment you trust do the same. Never read too much into a resume and never underestimate life experience or what your gut is telling you. Use it all.

After that, people need some direction. They need a coach to call the plays. They need a mentor to teach them the tricks of the trade. At times, they need someone to motivate them so they can achieve their very best.

Sometimes, even the best of employees lose their focus. They need direction and they need to be held accountable for what they have or have not done.

I've always felt managers who failed to do this were not managers at all. Maybe you've been fired by someone who failed to do his job first. I never really bought into the rolls-downhill style of management, preferring more of a buck-stops-here approach. However, sometimes you need to fire people. But as I mentioned earlier, discipline has its limitations and cost.

My dad's philosophy of management was "Don't manage with a big stick." You don't have to bully folks to get performance.

He's right. The stick approach just plain doesn't work well. The problem with this approach is that at some point you'll have to put the stick down to attend to other business, and that's when someone is likely to pick it up and smack you across the back of the head.

Leaders often need members of the team to surpass their individual goals in order for the entire team to succeed. We need them to be selfless for the good of all, and sometimes we need them to perform at levels they don't even think they're capable of.

Not likely for the manager who rules with a heavy hand. When the boss's job is on the line instead, he or she is more likely to see employees reach for the stick, under-perform, and exact a little revenge. Destroyed by the very monster they created. How poetic, Dr. Frankenstein.

There does, however, come a time when discipline and even termination is proper. When and how to properly do both is a matter of leadership.

At some point in your career as a manager, it is right and required that you fire. You have an obligation to your organization, your fellow employees, yourself, and even to those being terminated to do it.

I've been fired several times, and while there may be some disagreement as to how we got to that point, I know looking back it was the best thing for me. It forced me to grow. Out of that growth have come new careers, opportunities, and, in some part, even this book. So when employees ask to be fired, don't hesitate.

One wise friend put it this way: "Do not rule with an iron fist.   But every once in a while someone will stroll in your office, lay their head on the chopping block, and beg you to take it off. When that happens, don't hesitate."

At the end of the day, leadership is, in large part, about accountability. Everyone needs to know that at the end of the hall there is accountability.

Accountability is often met with a groan, but done well it's an absolute key to success. While we may not enjoy some of the attention it brings, the fact is, we all perform better with it.

It's important to note accountability does not mean nitpicking or micromanaging. Instead, it's establishing clear goals, communicating those goals, coaching, mentoring, and leading.

Once all that's done, it's holding everyone responsible for his or her contribution to the plan that's been laid out. You need to make yourself accountable too by making it clear what you'll take ownership of as well.

We should try every means necessary to help people grow and reach their goals, including the last resort, which is to let them go so they can pursue different opportunities—opportunities that may allow them to achieve their full potential.

THE POWER OF ONE

"It does not require a majority to prevail, but rather an irate, tireless minority keen to set brush fires in people's minds."
—Samuel Adams

In my travels I have often encountered what I like to call "The Fear of One." Why are people willing to tell me the right thing privately after a meeting but afraid to speak up or even show body language that would indicate their feelings during meetings?

A book on leadership would not be complete without acknowledging the fact that there are natural born leaders among us, those who, through a tremendous amount of self-confidence, don't need a consensus and who don't need to wait and see which way the room is going before they react. They demonstrate consistently the "Power of One," a defining characteristic of a true leader.

These leaders don't just tend to end up on the right side. They also nearly always have the ability to succinctly argue their position when their beliefs are challenged.

This group of natural leaders tends to be small. What do they have that the rest of us may lack? Sure, they seem to have a great deal of self-confidence, but it's more than that.

They seem to have no fear; fear that most of us seem shackled with. They are not afraid to stand alone. They understand the power of one.

Instead of the power of one, I often see something that looks like this: After meeting with a group of employees, I was alone, packing away my equipment, when a gentleman who had just attended the session walked nervously back into the room and approached me.

As he shook my hand, he thanked me for the message I had just delivered to him and his peers. He told me it was right-on and that he had no doubt it would do them all a bit of good if they would just listen and implement my suggestions.

All the while, he was constantly looking over both shoulders to make sure none of his fellow employees caught him talking to me.

The monster of peer pressure was so prevalent in this company that he dared not be caught talking to me, let alone encouraging me. The fear of standing alone was too much for him to risk with his fellow employees. Fear is a powerful monster—and it's one of the easiest to feed.

There is, of course, the other side of this coin: those who would rather stand completely alone for what they believe is right than join a whole army they thought was wrong.

Working once in a 2,400-member steelworker environment, I met a leader who embodied the Power of One and it made him an impressive leader, to say the least.

I had a group to work with over a two-day event. During the first day, one of the union participants began to inquire about our quitting time, which was scheduled for the end of their workday at four that afternoon.

The gentleman pointed out the bargaining agreement called for their release at 3:45. He explained that the men were allowed "off the job early" to complete a decontamination process. I had them for the entire day; therefore, there was no need for this process. I was told to hold them until four. My guess was I was being drawn into company politics.

In short, he wanted out early simply because the agreement called for it.

After double-checking my facts with management, I informed him we were indeed staying for the full scheduled time. Throughout the day he continued, with quite a bit of encouragement from others, to needle me for an early release time.

During one of our discussions, a gentleman who was sitting in the back of the room asked if he could add something to the conversation. Here's what he said.

"Every eighteen months we kill someone in this plant." That was a staggering statement by itself, but he had more to say. "You all know we have tried to help Jimmy's family as much as possible," he said. (Jimmy, I surmised, was the latest victim.)

"We've held fundraisers and made sure we were there to coach his son's baseball team," he said, "but our efforts, of course, will never replace him or what his family has lost. I've sat here and listened to our guest. I've heard nothing but information that we could all use to protect us and our families, and until I hear something contrary to that, I believe we should give this man the benefit of the doubt and hear him out."

He then apologized to me for taking up some of my time and turned the group back over to me.

Wow, the power of one! In an environment where it was not only okay to be disruptive but was encouraged by others, he was not afraid to buck the trend and speak up for what he believed in.

I did not have a chance to thank him for his assistance. When the meeting broke up, he left immediately. I asked one of the managers in the room who that was who had spoken up.

"He's our union president. Impressive, huh?" the manager told me. I'll say. He felt his people needed to hear the safety message I was there to deliver, and, with his leadership, he made sure everyone did.

You might be saying, yeah, but he's the union president. Of course he stood up. I can't tell you how many times I've watched so-called leaders, union or not, let the temperature of the room dictate their actions.

During the second day of training, I was again quizzed by the same employee about when we would get out. This time it was with a bit of a playful tone. I said four. He chuckled and said, well, technically, it was a violation of the agreement and he was tempted to file a grievance.

Right on cue, the president chimed in. "Go ahead," he said. "By the time it hits my desk, this fellow will be long gone." That got a good laugh from everyone and we went back to work.

Don't get me wrong. A variation—the Fear of One—can be a good thing too. As one consultant I admire greatly put it, "When making big decisions, organizations tend to fire individuals, not entire groups." He used this philosophy to build consensus among organizations in large revenue sales opportunities.

I love the philosophy when allies are needed to get something off the ground. But the Fear of One can morph into its own little monster if an environment where someone is afraid to question the group is created.

This fear is likely there because the culture does not support stepping out of the pack. Instead, they are expected to conform to the system and toe the company line or bow to "group think."

Here's the problem: some of the best ideas on how to improve all aspects of your business are not being heard. If your culture does not afford individuals the right to be heard respectfully, then you won't likely hear from them at all.

Often what happens is that someone comes up with an idea about how to improve things or challenges conventional wisdom but the idea is shot down immediately by a manager who either perceives the other as not being a team player or, in some cases, as a challenge to his position. It is that much worse when it's done by someone in higher leadership position.

If you want to hear the very best ideas on how to improve things, change the climate of the organization. Create a climate where open and frank discussion is encouraged.

As long as accountability and respect are present, you'll be amazed at how much disagreeing can improve things. If not, The Fear of One Monster will sit by silently gloating, and some fairly good ideas that could have improved your organization will go unvoiced.

WE'VE GOTTA TALK

"Communication works for those who work at it."
—John Powell

Perhaps the most effective way to kill any monster is through open communication.

Recently, while visiting with a client, I had the pleasure of seeing a Master Monster Slayer in action under a most difficult situation.

The CEO of this mid-size company called an all-hands meeting. The recent skyrocketing fuel prices were really taking a toll on their business. Transportation costs were bad enough, but the commodities prices, which made up the bulk of their operating costs, were soaring out of sight.

All of this had produced a shortfall of tens of millions of dollars. As a result, some swift and decisive leadership was called for. She opened the session by reminding everyone of her commitment to the idea of transparency and candor through good times and bad.

She told them shortfall would mean layoffs, which would be

announced at the end of the day. Some locations, she added, would actually be shut down. This painful move was needed to ensure that the company remained financially healthy and positioned well in the market place.

She called on all who would survive the layoff to pull harder and to remain focused through what was shaping up to be the most difficult year in the company's nearly ninety-year history.

No one wants to deliver bad news. It's a lousy job, but she understood she could either dance around it and dress it up or meet it head on.

The rumor mill was taken offline before it had a chance to gather steam. I have no idea how many potential monsters were slain that day, but the company is stronger today because the very top of the company understood that you can't feed the monsters in good or bad times.

Transparency and candor at all times have become choice weapons of Master Monster Slayers.

WE GOTTA TALK SOME MORE

Good things just seem to happen when people communicate.

We often get off track only when we quit communicating or quit doing it well.

In 1974, TWA flight 514 crashed on approach to Dulles International Airport near Washington DC, killing all eighty-five passengers and seven crew members.

The investigation uncovered that a pilot for another airline had a near miss under similar circumstances months earlier. The problem was that the near miss was only communicated to the first pilot's airline.

As a result, the Aviation Safety Reporting System (ASRS) was created to ensure such near misses are reported across the industry so that all airlines benefit from such information.

ASRS has been collecting confidential, voluntary reports from pilots, flight attendants, and air traffic controllers since 1976.

Also of interest is that NASA, not the FAA, has oversight of ASRS. NASA, while respected in the aviation community, has no governing authority over commercial aviation, and therefore can in no way "come back" on any one who reports to them.

It's one thing to encourage open communication. It's another thing to create an environment to ensure it happens. So does all this communication work? Since its inception twenty years ago, fatal crashes have fallen 65 percent, down to one fatal crash per 4.5 million departures.

That's not the only effort in the aviation industry to improve communication.

Crew (or Cockpit) Resource Management (CRM) training has also been deployed to improve air safety. Research found that the primary cause of the majority of aviation accidents was human error: mainly interpersonal communication, leadership, and decision making on the flight deck.

Part of CRM is to create a culture where respectfully questioning authority is encouraged. Cockpit recordings of several air disasters revealed first officers trying to bring critical information to the captain's attention but in an indirect way.

Tragically, many times this information could have been used to avoid disaster. CRM, which was developed out of a 1979 NASA workshop, does much more than work to improve communication in an environment where, historically, what the boss (captain) says goes. But getting crews talking effectively may be its biggest achievement in improving air safety for us all.

Following September 11, the airline industry went through some tough economic times. But once consumer confidence in security returned, so did consumers. Fact is, we are moving more people safely in North America through commercial aviation than we ever have thanks largely to better one-on-one communication.

Open, honest, respectful communication must be the goal. Get folks talking and create an environment where it's not only okay to ask questions, but where respectful challenges are sought out and encouraged. If you're not being challenged, then you're not receiving the nourishment your organization needs. Instead, the monsters may be hording it for their own indulgence.

THE PIGEONHOLE

"The way you see them is the way you treat them, and the way you treat them is the way they often become."
—Zig Ziglar

When you get the opportunity to work with various people from different walks of life, you start to notice some basic human tendencies. There are things we all have in common.

For instance, rushing, frustration, fatigue, and complacency cause just about all accidents or mistakes, regardless of where they happen.

Give it some thought. Can you think of a time when you hurt yourself when you were not in one or more of those states? Trust me here. Those four words in part have made *SafeStart* author, Larry Wilson, a pile of money.

There are many things humans share regardless of culture. Abraham Maslow in the 1940s postulated his hierarchy of human needs. The first and most basic need for all humans is physiological. Obviously, if you're starving, who cares about self-actualization or "being all you can be," which was the fifth and highest need in Maslow's hierarchy.

I've noticed another need we all seem to have, and that is a need to pigeonhole people.

The general idea is to determine as soon as you can who someone is. Whether your assessment is accurate or not is not of any consequence when you're doing this. Just put them in their place as soon as possible so you don't have to deal with them on any real level. It's unfair, very calculating, and very, very common.

When we pigeonhole folks, we limit their opportunity to grow. We cut off developmental tools, encouragement, and the will to grow. And when they fail to be all they can be, we reserve the right to say, "See, I told you they would never amount to anything." It becomes a self-fulfilling prophecy and a way of controlling others.

It seems simpler when we can fit people into little boxes. It makes it easier for us to deal with them. I think it has something to do with our basic need to control our environment. If we already know or think we know who people are, then we can assume how they will react and what they will do.

A psychologist who has developed a method for improving organizational performance by focusing on these types of issues related a story that illustrates this point.

One day, while standing in line at the grocery store, the psychologist noticed a mom and her four-year-old daughter.

As he made eye contact with the little girl, she smiled and waved. He acknowledged her by asking, "How are you, young lady?" Mom quickly ended the exchange by turning her daughter's little head into her thigh and saying, "Excuse her, she's my shy one."

What do you think the odds are that that little girl will grow up to be shy?

Now, there's nothing wrong with being shy, except that it almost always guarantees that you will not be a strong, effective leader. It's just a fact that extroverts tend to get ahead in the world of leadership. That little girl was being strapped with a handicap of sorts that would likely haunt her career and her life forever.

But that's what we do when we pigeonhole. We're contributing to the programming of that person. We're limiting who they think they are and what their own perceived value is.

It might not be a bad idea to remember that personalities are often, in part, programmed at an early age, and that strong leadership may even be able to help a person break out of her personal chains and find success that had previously been out of reach.

Leadership is, after all, about helping others reach their potential.

Most of this pigeonholing likely comes from another less-than-attractive human trait, which is to only seek information that confirms what we think we already know.

Most of us think we have a decent handle on how things work. We then set out seeking information that confirms what we already know, and we often dismiss any evidence that would suggest anything to the contrary.

Over the years, I've trained thousands of individuals to lead processes for co-workers, and I've been amazed at by how many adults are afraid of public speaking. Some are downright terrified.

I was working with a group once and the training called for them, after a day of instruction from me, to stand up and deliver about five minutes of material that had been prepared for them, along with a personal anecdote relating to the topic.

A gentleman in my session, looking rather nervous, approached me at the end of the first day after everyone had left the room. He was concerned about his ability to deliver the material to the group.

He told me that public speaking was his biggest fear. With a lot of effort on his part, he had managed over the years to get up enough nerve so that he could handle talking to thirteen people, but no more. He actually knew exactly how many people he could speak to without succumbing to fright.

We had a significant problem. Counting me, there would be sixteen people in the room the next day. He had already completed the headcount in advance and was sure he would not be able to deliver on his assignment. I assured him that no one expected perfection and asked that he just give it a try.

The next morning he showed up feeling worse than before. Just thinking about the exercise made him physically sick. After throwing up, he went home for the day. He never even made it to the front of the class.

What a terrible burden to have to carry, and I wonder where it all began. Was he programmed to be shy? Was he told he was Mama's shy one? I'll never know, but I do know his chance of becoming a great leader is virtually non-existent.

It doesn't matter how smart or technically proficient someone is. The demand for leaders who can't stand up and talk in front of their own teams is limited, at best.

Anyone who does not conquer this fear on his own or with the help of others will remain forever trapped as a mid-level manager, regardless of his talents. Worse, the limited world his own inhibitions have forced him to inhabit represents a new breeding ground for monsters. Inevitably, he will resent that peers are being promoted while he remains in the same position, year after year.

It would be disingenuous to say that I was a shy child, but it would be fair to say that I was terrified of public speaking. A public speaking course helped tremendously, but there was still a lot of work to be done.

Over the years I have given hundreds of training sessions, spoken at many conferences, given keynote speeches for organizations and corporate functions, and been hired as a motivational speaker.

Early on in my career, a wonderful leader of mine made it clear I would not succeed in any organization until I mastered the ability to speak to large groups.

He helped me out of my pigeonhole. I'm not sure just how I got there, but I know how I got out: with good leadership and a lot of work.

Who's being held back in your organization? What kind of talents were lost years ago when someone inadvertently locked them away? What are you doing as a leader to rescue some of your own people and to help them be more productive and successful?

Avoid the temptation to pigeonhole people and help release that untapped talent your organization possesses, and in doing so, you may prevent some monsters from even being born.

ROI (RETURN ON INTELLECTUAL CAPITAL)

"A society grows great when old men plant trees whose shade they know they shall never sit in."
—Greek Proverb

Mentoring is going on in every organization. The problem is that most of it is informal, and it's almost always negative.

It goes like this. New kid shows up on the job, excited about what the future has to offer. He or she almost immediately looks for someone to show him the ropes, and the moment that connection is made, our mentoring process is underway.

Far too often, it begins this way: "While I appreciate your excitement and all, kid," the old hand expounds, "I promise you'll soon be beat down like the rest of us."

Sound familiar? Now set aside the fact that in some cases the so-called scoop on the company is dead on. That clearly needs to be addressed. But I'm talking about the naysayer capturing a new hire and recruiting him to join the monsters, those people who are sucking the very life out of the organization.

An obvious danger of informal mentoring is that you lose complete control of the process and you may not get any of the benefits formal mentoring brings. And that may well be the best-case scenario.

We've got Bill, for example, who is one of our most knowledgeable maintenance guys. Bill is sixty-one and already has a running calendar of how many days until retirement.

Keep in mind he's planning to work until he's sixty-five. While Bill is certainly good at what he does, he would hardly be described as a pleasant fellow. Bill is more of a stay-out-of-his-line-of-fire guy.

Years ago, Bill and the organization came to an informal agreement. No one ever talked about it, but both sides agreed in essence that they would leave the other alone and try to coast to this retirement thing with minimal friction.

So Bill plods along, counting the days until there are no more and he can walk out the door with a gold watch and a hardy, "So long, suckers."

Here's the problem. Bill also walks out with a tremendous amount of intellectual capital. You have trained Bill. You paid him all those years, and now he's walking out the door with all that intellectual capital.

You sent him to school and invested untold resources in his training. The company was with him every step of the way, but now he'll be going and you have to replace him.

Fact is, you can't replace his experience or the intimate knowledge that he has of his job and your business. So he walks and you're left trying to fill the void.

Actually, many folks see this coming a mile a way and take action, sometimes as early as five years in advance, by hiring a young maintenance person. Let's call him Billy. Billy is expected to kind of grow up on the job pretty much on his own. When Bill does retire, there is a tremendous step down in talent, not to mention experience.

Often this is exacerbated by Bill's notion of "showing the kid the ropes," which amounts mostly to handing over only insignificant or grunt jobs and never really teaching him anything.

Or worse yet, showing the young pup the "tricks of the trade," which may not be what you want Billy to learn at all.

For example, Bill long ago quit locking out (de-energizing) certain equipment to work on it. It's an outright safety violation, but Bill knows what he's doing (or at least thinks he can manage it without following procedures), and besides, he could see an accident coming a million miles away. He's done it that way for years and really doesn't even think about the potential danger anymore.

When he's showing Billy how to do it, he shows him the way he's "always" done it. The problem is that Billy does not have the experience to see the accident coming and goes about doing the job the way he was trained. He's a sitting duck.

Later, maybe weeks, months, or even years down the line, there's an accident. Billy gets hurt or worse, and the cause of the accident is written off as Billy not following procedures. In fact, he was trained not to follow procedures.

Years ago, unions by and large took care of mentoring for most organizations. It was through formalized apprenticeship programs that we found our next generation of skilled craft workers.

The influence of unions, for better or worse, continues to erode in our workplaces, and so we continue to lose another very important leadership incubator.

There is a major shortage of skilled craft persons in our society. There are already far more jobs than skilled men and women to fill them. By the way, this organizational drain does not just apply to craft skills but to all skills that top performing organizations need.

Many are turning to a formalized mentoring process to help offset this loss of intellectual capital.

Formalized mentoring allows you to make sure the entire process, like any other, is meeting your expectations. Formalized mentoring should involve a selection process that establishes defined characteristics of a good mentor, proper training for that mentor, and support and concrete expectations for all involved.

It should be a joint effort between the organization and the mentor to make sure best practices are passed along to the mentee.

Let's go back to the case of our two Bills. The elder wants out, the younger wants to get ahead. Take those final five years of Senior's career and pair him up with Billy as a formal mentor with very specific guidelines and training on how this process will evolve.

If Bill is placed in a formal role as a mentor, you're much more likely to get his best effort. It's a bit of his legacy if you will. Bill passes along the right information to Billy, coaches, encourages, corrects, and at the end of five years a tremendous amount of his intellectual capital has been transferred to Billy.

Bill's final years have been productive and energized with a new purpose and we are left with a young maintenance man who is wise beyond his years.

Fact is, Bill is leaving. What he leaves with in terms of intellectual capital is entirely up to leadership.

Formalized mentoring is becoming a key weapon in slaying monsters. Those who do it well are, in effect, installing leadership at all levels of the organization by transferring solid leadership traits to the next generation of leaders.

However, if you don't have a mentorship program, you can still mentor the new hire yourself or make sure you introduce him to the person you want to be a mentor to him. But, whatever you do, don't let the monsters get him first.

CAVE People

"Anger is a killing thing; it kills the man who angers, for each rage leaves him less than he had before—it takes something from him."
—Louis L'Amour

CAVE People are Citizens Against Virtually Everything and every organization has them. These folks are just never happy. If they won the lottery, they are the kind of people who would soon be complaining about the tax headache.

One of the toughest jobs I've had is trying to "flip" these people, so to speak. That term is what those hiring me use. I have never really warmed up to it in part because I never set out to flip a Highness of the CAVE People into a cheerleader.

By the way, I find the term flipping a bit simplistic because it is hardly an accurate description of what has to be accomplished. It's really more of a migration over time. I also find it unrealistic to suggest that I have the talent or possess any special persuasion techniques to flip anyone.

I just want them back in play, and if I can get them to quit actively fighting the organization, I call it a significant win.

Moving these people to a more productive role in the

organization pays huge dividends when it can be done.

If you performed a Pareto analysis on your organization, you would find, for the most part, that only a critical few are holding back the entire organization.

You may know the Pareto principle as the 80/20 rule. Only a critical few are responsible for the critical mass.

For example, about 20 percent of a sales group generates 80 percent of the revenue for the entire team. The rule also works in reverse. Typically only a handful of low-end performers create the vast majority of challenges for leaders.

The low-end performers are those I've dubbed CAVE People.

Ignoring these CAVE People is something I see many organizations try, but to do so is to ignore a significant opportunity to improve things, besides, not fixing the problem is feeding the monsters. But that's just what many managers do by tuning out, marginalizing, or just outright ignoring these folks.

The first thing I try to figure out is why in the heck CAVE People fight their organization so much. Often it's bad history, company politics, or even issues in their personal lives—things they find difficult to let go.

So, getting them past some of those issues may not be an option. My hope then is at least to sway them from actively fighting management.

It's even better if I can get these folks on my side and help them become process champions. If that happens, there's a chance of revitalizing a teammate, which is a win for everyone.

Okay, so the plan is to get the very people who drag the team down to somehow become the people who drive down the field for the winning touchdown as the clock runs out.

Again with the drama, but some of these folks are tough cookies. And when I have won big with them, I've had a tremendous feeling of accomplishment, especially because some in their organizations were amazed at what I was able to do.

My approach was to not let the past block a more productive future for us all. That was easy as an outsider. I also had to discount the negative reports that I received from others. Notice I said discount—I did not say ignore. There was probably valuable information in what I was being told by others, but that information was also colored by their bias and personal history.

Even if the information I was given about a CAVE Person was accurate, it was their history and not necessarily their future. All I could do was make sure it did not affect my perception of them or the current situation. If they had a troubled past, that had nothing to do with me.

Sometimes letting go of past hurts is a difficult chore, but it has to be done, or at least set aside for the moment if progress is to be made.

For my part, it was simple—I just didn't get caught up in the past.

The other factor that comes into play here is that CAVE People are often agitated by the very people who claim they want them out of the cave in the first place. Sound familiar?

We're back to pigeonholing again. Those who would gain by discounting CAVE People sometimes go out of their way to keep these folks stirred up.

When the citizens of the cave overreact as they are prone to do, it's easy to point fingers and say, "Look at those hot-headed, overly emotional jerks! We really can't take anything they say seriously."

Becomes quite the nice self-fulfilling prophecy: make sure they continue to fail so we can continue to point out their failures. Here we go again feeding the monster.

Often those perceived as disgruntled employees are those who simply gave up on an organization that long ago gave up on them.

Leaders who give up are failing us all and, worse yet, allowing disgruntled workers to become a drain on more motivated employees and the entire organization.

As an outsider, I want to identify and meet with the CAVE People, and I always find it interesting that I've never had a shortage of nominees of whom I should meet with.

In an initial meeting, I ask for a CAVE Person's input as a leader. Cave People are, in fact, leaders, just not the kind that are as helpful or as productive as they could be.

I try to find out just what their beef is. Keep in mind, I can't fix any of their problems. I don't have the time or the authority. In most cases, if I acted like I could or would take on their problems, I would have been rightly labeled a company rat.

My job is to get some input and hear their take on how things are going and what can be done to make it better.

My approach is simple: honesty and respect. To begin with, I really listen to what they have to say. Over the years I've learned a lot by listening, including the fact that listening is more than just waiting for your turn to talk.

Most Cave People are taken aback by my lack of comments. It's as if they're waiting for me to cut them off and tell them they're wrong. By the way, from time to time, that's just what I'm thinking, but it's not about who's right. It's about letting them get it out and then identifying how we get them back in the game.

So I listen and keep my judgments to myself and my mouth shut! If all goes well, I can help them and the organization funnel their leadership skills and talents back into a more productive role.

Early on in my career, I developed a reputation as a consultant who could work in organizations where things were not going well. It seems the more difficult the climate; the more I was called on.

One client I had was a forest products group. To say they were industry leaders in their business would be a bit of an oversell. They were not the biggest, but they were a substantial player in a very tight business.

Their largest challenge was that their past was clearly blocking their future.

They had recently changed ownership groups and top-level management. Their new CEO came over from the largest forest products business in the world. He was the only CEO in ten-plus years to take the time to personally sit through one of my entire sessions.

His commitment was real, not only in his communication but in his actions as well. Things were changing fast and he was driving it all.

Unfortunately, there had been many sins committed by the previous ownership group and the new boss had to pay for them. To his credit, he accepted that responsibility. He knew twenty-five years of mismanagement by the previous group would take time and a tremendous amount of effort to overcome.

He hired me to help with his change agenda. During the break of one of the many meetings I held with his employees I was met by a man who seemed a bit agitated.

He invited me out into the parking lot to talk. It was the first of a few meetings I would have with Murray, and while our time together was brief, he would become quite an influential person in my career.

Once we got outside, it was quite clear Murray was upset. He quickly pointed out that mismanagement was creating not only accidents but a myriad of performance problems.

I agreed completely. After working with hundreds of companies, I had yet to find one organization that was not guilty of this to some extent. So the company needed to do a better job managing its people and resources.

That certainly would help us keep people safer, more productive and so on. But I also had a question for him.

There are nearly three times as many injuries and ten times as many accidental deaths off the job in the United States and Canada, so who's creating the risks off the job? The company has little if any control there.

He answered that the individuals themselves for the most part and occasionally other people around them created the risks. Again, he was correct.

I thanked him for his frankness and told him I've always believed you only have a right to point the finger at the organization if you are willing to accept your role in its failure. Blaming the boss for off-the-job mishaps just doesn't compute.

Bottom line: the company had to get better, and everyone involved had a role to play if things were going to improve.

It doesn't matter where you get hurt; it's almost always about behavior. In short, it's all of us together that cause the problem.

So it seemed Murray and I we're seeing things the same way. That wasn't surprising to me, but I think it caught him a little off guard. Murray, I assumed, would take the position that his behavior needed to change. End of story.

It would be tempting to expect that I would suggest the company's behavior was fine. After all, they were paying me to speak to employees about their behavior, right?

Murray's point was a good one and I planned, with his permission, to share it with the entire group when the meeting resumed.

As I walked back into the auditorium, I was quickly approached by the site safety manager who asked me if I had been involved in a conversation with "that man." He was pointing at Murray and there was more than a bit of animosity in his voice.

I said, sure, he had some questions. The manager's response was immediate and hostile. He said, "Do not talk to him. He is the most negative person in the entire company. In fact, he is a cancer on this organization." When you're described as organizational cancer, that cannot be good.

"Actually," I said, "he made a good point. The company is sometimes responsible for some of these situations and my experience tells me he's right."

"I knew it," the site safety manager said. "I knew he would get in the middle of this and foul things up."

Our conversation went back and forth, him trying to convince me that Murray was the embodiment of evil while I maintained he seemed like a decent enough fellow.

In the end, I decided to invite Murray to our leadership group session. It was this group that had the primary responsibility to roll out the new process to their fellow associates.

My recommendation was met with a firestorm of controversy, and there was plenty of advice to abandon the guy before he took us all down with him.

The good thing about being the so-called expert is that you hold the trump card. After all, they were paying for my opinion, so they had to allow for an outside chance that I might be right.

Therefore I could invite anyone I wanted, and I made it clear that their decision not to originally include the union, in which Murray held a leadership position, was a mistake. Here was an opportunity to make it right.

I won the political battle to involve Murray, now I had to win with Murray.

I invited him as an observer to a series of meetings that would last two days. I made it clear I expected nothing out of him for attending. He could participate as much or as little as he thought necessary.

I also made it clear that I felt overlooking the union's input on this project was a huge mistake on the company's part and that he should join us if for no other reason than to answer his and the union's concerns. He agreed and we were underway.

Murray spent most of the first day quietly taking it all in. He participated in everything I asked and was clearly giving the new safety approach meaningful consideration.

I wasn't sure what to expect, but I have great faith in the

process. I felt that if I did my job right, Murray and I could easily close the perceived gap between him and the others and he could become a strong leader.

At the end of the first day, Murray asked if he could address the entire group.

Here's what he had to say:

"After three decades of dropping trees with chainsaws, I have learned a few things. There is no better safety tool than the faller [someone who cuts down trees with a chainsaw] himself.

"It is impossible to engineer the forest. There are sinkholes from rotting timber (those are actually covered with snow in the winter, so good luck seeing them); there is weather to contend with, the actions of others, the list of hazards is endless.

"A good hand-faller must have a working knowledge of physics, math, and good old common sense. Dropping a tree is part science and part art. To watch a skilled faller perform is truly an amazing thing to see.

"But at the end of the day, the only sure way to make it out of the bush and home to his family lies completely with the faller himself. We must be aware of the hazards at all times. All the rules, policies, and procedures do you no good if you don't have your eye on the ball and your head in the game.

[The safety process I was there to present focused on awareness and the mental states that cause distraction.]

"I have not seen a tool this valuable for our guys in over thirty years in the bush. I cannot for the life of me figure out why any seasoned faller would not see this as an absolute benefit both on and off the job."

I was inspired by his eloquent words. The rest of the room was in shock. Here was their "cancer" not only supporting this initiative but standing up publicly and assuming a leadership role. It was quite a moment.

To have Murray speak so glowingly of the safety process was

a high-water mark for my career. I told him I wish I had recorded the moment because the marketing department of the company I worked for would run with such a statement from a union leader.

Some unions traditionally have not supported processes that look at behavior as a cause of accidents.

Murray went one better. The next morning he handed the same statement typed up with his signature and the number of his union local. I cautioned him that I would give it to the marketing department and they would indeed run with it. His name would hit just about every leader's desk in North America, union or not.

"I stand by every word," he said. "You do with it as you wish."

Think back to earlier when we discussed the CAVE People concept. Murray was never going to be the cheerleader the company wanted. There was too much bad history between them and it wasn't his nature.

What they tried to do was isolate him. My guess is they thought they were "controlling" him by not listening to him. Instead, he quickly migrated to his natural strength as a leader. The problem was he was leading the troops against the organization.

I would have paid top dollar to see Murray walk back into camp later that week and hear his guys greet him with, "New program same old BS from corporate, right Murray?" I'll bet his response was just as shocking to his men as it was to those in our meeting.

I would like to think that our chance meeting helped Murray get back to what he's really good at: leading tough men to a better overall work environment. I've thought a lot about Murray over the years. Trust me, I did not "flip" him as some suggested. He simply took me up on my offer to lead.

The company cancer was actually an inspirational leader waiting to be released. I was asked by many in that company how I managed to flip Murray. As the years have gone by, I'm even more convinced that there was no such flip.

We simply listened respectfully to each other and, in the end, our process made sense to him. I dare say if I did anything for Murray it was to simply show him respect long denied by others. That opened up communication between us. Lack of respect, which seems to be everywhere, is a smorgasbord for the monsters.

Murray became a process champion and a lot of our success was because of his leadership.

Look, Murray had a responsibility to say things in a way that would increase the likelihood he would be heard. But an essential part of leadership is to hear what you need to hear, especially when the person is not saying it well.

Good Monster Slayers know the difference between bitching and the ideas they need to hear.

Before you banish someone to the cave, hear him out. Ask if he would support your initiative if it made sense to him. If so, would he give it his public seal of approval and join you in supporting the initiative? His answers just might surprise you.

So-called CAVE People are often just frustrated natural leaders. Good leaders figure that out. Go in and get them out of the cave!

MONSTERS OF THE DEEP

"The lady doth protest too much, methinks."
—William Shakespeare

I've been amazed at what we are willing to do in organizations to placate bottom dwellers, the Monsters of the Deep. You know the type—they complain the most, produce the least, and always seem to be grinding a political ax.

It's not like you need the 80/20 rule to find out who these people are. They seem to enjoy and thrive on creating chaos.

Through political correctness, politics, laziness, and outright fear of being sued, we allow these critical few to drive all of us right into the ground.

Let's be clear—every human deserves to be treated with respect, but it seems to me we have lost ourselves in political correctness and along the way found a new way to feed the monster.

All people are entitled to respect. It's their birthright as humans. Now, of course, some grow and earn far beyond that initial respect while others squander their own and tear down others. But that's an individual choice; everyone should start off equal. Where we go from there largely depends on how much respect both sides decide to grant or deny one another.

What politically correct has become in many cases is the Bottom Dweller Monster's weapon of choice. They sit back and listen, always scanning, and if they hear anything that can be remotely twisted as offensive, they run down to human resources, beating their political drum all the way.

This lesson was driven home to me once when I was working with an especially tough group of employees.

To say these employees were angry would be like saying Albert Einstein was pretty good with math.

This was a tough group, and they had a right to be angry. Many of them worked under some of the most awful managers I have ever seen. This organization, with only a few exceptions, lacked respect for individuals at all levels.

As the months of training dragged on, I endured abuse that I frankly expect would have gotten me fired had I been the one dishing it out.

I was told repeatedly to go F*** myself and often referred to in person as a son of a B****!

On one occasion, a "lady" accomplished both in just a single training session.

Tough job, but when you enter this line of work, it comes with the territory. This was just much worse than most.

One day after a particularly rough meeting, I was approached by the human resources manager, who notified me that there had been a formal complaint made about the presentation I had wrapped up only minutes earlier. Wow! They didn't waste any time!

Despite knowing I hadn't said anything that would be considered offensive from an HR perspective, I immediately backed up the tape in my head and tried to figure out what I could have possibly said to offend someone. I was at a loss.

The HR manager informed me that I had referred to the entire room as "guys," as in, "Guys, in this case you need to be looking out for (whatever I was talking about)."

This so-called lady—the only female in the room—found the term "guys" offensive. This was the same woman who pulled the rare exacta of personal assault just moments earlier. Have I made it clear she was very explicit in her choice of words?

I thought it interesting that I was being called on the carpet for the word "guys" in light of what she had actually said to me. Being an outsider, I know it's not HR's job to provide me protection at any level, but her complaint was really kind of silly when you think about it..

Someone who was so willing to viciously attack me verbally was going to draw the line of decency at "guys"?

Okay, it's not that silly because she had an agenda and I had given her an opportunity. She hated the company she worked for, and I'm sure she didn't think much of me either. I was given a verbal warning from human resources and asked to make a formal apology.

Later, the complainer's boss told me that my offended nemesis was a constant underperformer who was always stirring things up and that I shouldn't take it personally.

As she told me, "I have to put up with this kind of stuff every day."

I wonder about people who hate that way; the energy to maintain that level of anger must be exhausting.

Bottom line: she had taken a system that had been installed with the best of intentions to ensure that all employees were afforded a respectful place to work and used it against not only me but her own organization.

She was making a mockery of the system, the organization, and the very real victims of hate.

Instead of having her complaint dismissed for lack of merit, she was being rewarded and indeed fed by the HR manager, who decided to cross the legal Ts and dot the Is out of fear that failing to capitulate would rebound on her or the organization. That's called extortion.

HR wanted an apology from me, hoping it might make this all go away, at least this time.

I gave apologizing some thought and declined, explaining I would not feed their monster for them. My client smiled and said, "You're right. I'll let HR know your decision." I never heard another word about the complaint.

Over the years I've learned to let such attacks just roll off my back. After all, they're not really about me. I wish I could say this was an isolated event, but most companies I've been involved with (and it has been hundreds at this point) seem to have developed this relatively new breed of monster to one degree or another.

The hate that some people have and the amount of energy and disruption they will inflict on others is astounding.

But hate if channeled properly can be a good thing.

Dr. Henry Cloud, in his book *9 Things You Simply Must Do*, put it this way: "Hate well! Tell me what you hate and I can tell you who you are."

Well, I hate, among other things, those bottom feeders who continually underperform and yet demand to share in the wealth of others' successes or just seem to be obsessed with stirring the pot of discontent versus getting something constructive done.

As a much younger man, I worked for a manager who didn't care much for bottom feeders either. In fact, he made it a point to starve them out. You could call him a modern day Robin Hood in reverse. While the legendary woodsman stole from the rich and gave to the poor (actually he stole from the corrupt tax collector and gave it back to those who had earned it, but that's another story), my manager took a slightly different and equally noble approach.

First of all, he felt the need to steal from no one, especially in the name of being fair. Low-end performers made little or no money over their base. There was no reward for staying at the bottom other than just holding a job for the moment. All the reward (mostly money) was shared by that small group of top performers.

This manager's idea was to reward the top level of his team to keep them motivated to continue their success.

But he was also sending a clear message to those in the middle of the pack, which happened to be the majority of the team— they were just a bit more effort away from joining the elite, well-compensated group.

Everyone seemed to be fine with his approach with the exception of the bottom feeders. They classified the plan as an attempt of the manager to take care of his buddies. I always thought it was funny that his so-called buddies were only the consistent top performers and those clearly on the rise.

The bottom feeders complained it was simply unfair and they wanted their fair share. But were they entitled to it?

They brought little to the overall team in terms of revenue, while at the same time costing the organization the same amount of money to employ—hardly equal value.

In fact, the lower-performing employees often took up far more time and other resources from the leadership group and actually cost more to employ.

In this particular organization, these bottom feeders came and went as if they were being fed to us through a revolving door.

Once in a while, a top performer candidate would pop out of this group. He or she was quickly identified as having some potential and received attention from the leadership group, whose job it was to develop the next top performers.

Often this management approach is dismissed as unfair. What's really unfair is asking everyone to carry the load of the bottom dwellers and then compensating them with what others have earned.

It is unfair to the top performers who do more than their part and it is unfair to the company that could get a better return on its investment with a more productive employee.

It is not fair that most employees have to be subjected to verbal abuse and exposed to someone who is always negative.

Many are starting to look at employees directly for what they bring or don't bring to the organization in terms of talents, attitude, and desire. Jack Welch, in his book *Winning*, calls it differentiation. There are many arguments against this type of approach and, as Jack says, "They all amount to excuses."

When seeking out monsters, look below; many live at the bottom just looking to make everyone as miserable as they are while swimming by with as little effort as possible.

THE TIMES THEY'RE A CHANGING

*"There's a battle outside
And it is ragin'
It'll soon shake your windows
And rattle your walls
For the times they are a-changin'."*
—Bob Dylan

Not really sure what Bob was thinking when he penned those words, but change does shake windows and rattle walls. It's downright painful to change.

Change comes about when the pain of staying the same outweighs the pain of change, or so I was told once by a "Who Moved my Cheese?" coach.

If you're not feeling any organizational change pain at the moment, you better get busy. Organizations that fail to change are simply left behind. So the pain of change is inevitable and constant for those who are getting ahead.

Change not only comes with pain, it may even come with hate. If you want to know hate, introduce change. If you want to know both loathing and success, become a champion of change.

So what is it about change that gets employees rolling their eyes with that unmistakable here-we-go-again smirk?

Why are we so resistant to change? Change is supposed to be a good thing. We're growing as an organization and the change is designed to make us in some way better, more profitable, efficient, productive, or safer.

Well, for one thing, regardless of the motivator, change is painful. So the first question is fairly obvious. "Is the change we're about to make worth the short-term pain?"

I say obvious because many managers seem to be stuck in the change-for-change sake role.

More to the point, they seem to be mostly concerned with their own agendas or putting their own mark on the organization without giving much thought to the actual change before them or how their employees will perceive it and be affected by it.

I once worked for a sales group that went through three vice presidents of sales in less than a calendar year. Each new vice president felt the sales floor, which contained 400 folks, was badly structured and organized, and each one of them, shortly after his or her ascension, set out to rearrange the floor in an effort to improve profitability.

Now this was more than just moving bodies. Cubicles had to be packed up, moved, and unpacked. Phone lines had to be rerouted. Computer connections had to be reconfigured. It was no small undertaking.

Each move required two days just for the salespeople to pack up and move. Six days (three different VPs, three different versions of moving day) times 400 people equals 2,400 working days lost, not to mention the time and effort of the phone crew and IT department.

All this change, intended to bring about efficiency and profitability, instead created chaos and sent productivity crashing to the ground.

Now think about what all this wasted time did for the morale of a commissioned sales group that was taking it on the financial chin every time a new VP decided to send them all packing.

It was a giant game of adult musical chairs and no one was winning. Worse yet, new monsters were being created with each move as the sales team sat back and watched its commissions erode away along with its confidence in leadership.

Of course, management did not see or intend it to be that way. They intended the change to bring about a better work flow and, in turn, improve productivity.

What they failed to do was measure the risk of that change from a historical or futuristic perspective. There is something to be said for getting it right the first or at least the second or third time.

Leadership, especially at the corporate executive level, has become very fluid. Most people at the top today only occupy that position for a short while. Either they are wildly successful and leverage that into the next fortune 500 job or they are miserable failures and get shown the door.

Either way, they have a short amount of time to make their mark on the organization. Often a casualty of this quick turn around is that most of what was done before them is arbitrarily dismissed and replaced with the new approach. As far as they are concerned, a future beyond their tenure doesn't exist.

When this happens, it comes with a cost. One of the costs is that long-term employees realize that the "exciting changes" proposed by the new chief are probably not worth taking time to understand, let alone buy into.

After all, the new approach is going away when the next "new" boss arrives.

So employees see constant change with little or no consistency and conclude that anything new is quickly heading the way of the flavor of the month heap.

My father went to work for a company sweeping floors at

fifteen; by age forty-five he was one of the top executives.

Those days, sorry to report, are pretty much gone. Instead of having one or two jobs over our working lives, most of us will have closer to ten.

Heck, many of us today have had several careers, let alone jobs. I'm on four and counting.

With that kind of movement, positions at the top tend to be fluid. According to *Forbes,* the average tenure for fortune 500 CEOs has now slipped to a mere six years.

The company with the "moving-day" vice presidents had five different presidents over an eight-year period.

It's not likely that this fluid nature is going to change anytime soon, but how we factor that into our decision-making process could make a substantial difference in our careers and the careers of those we are paid to lead.

The bottom line, to quote rock icon Geddy Lee of RUSH, is "Changes aren't permanent, but change is."

If you want to stop feeding the Flavor-of-the-Month Monster, keep in mind that the window of opportunity to get the team on board when it comes to change is brief and starts closing the minute you announce the change.

Think it through, front load as much value for as many as you can, keep expectations reasonable, and work your tail off to sustain it until it has a chance to mature and everyone is in a better position to judge the value of the change.

Be part historian and part prophet. Recognize others have endured change before you and will continue enduring it after you're gone. So make it worth the effort.

Change is constant, but eliminate change for change sake. Celebrate the success that proper change brings and make sure you develop leaders and champions of proper change who will keep it going long after you or the next boss is gone.

HUNTING SUPER MONSTERS

"The truest characters of ignorance are vanity,
and pride and arrogance."
—Samuel Butler

You'll find monsters in the strangest places—even among top producers. It seems strange to look for those who are dragging you down among those who are performing at the very top, but sometimes being the very best can be intoxicating.

They've been constantly praised and have all the rewards and trappings that come with top performance, as they should. Sometimes, however, they begin to believe all that they are being told and begin to coast or, even worse, think they are indispensable.

The problem comes when their success goes to their heads. They become demanding and intolerant of their peers instead of helping them.

Unchecked, this little monster can quickly grow into arrogance and make life unbearable for everybody, including themselves.

Left on their own, these employees can begin to believe that the rules don't apply to them. In some cases, they will even decide for themselves what the rules actually are.

This monster can prove to be particularly difficult because leadership gives it a long leash and often would rather not deal with it.

On one hand, leadership wishes others would produce like that. On the other, they are concerned about the overall health of the team.

It's this lack of action that often causes things to come to a head.

Look, we need our top producers, but at what cost? Top producers tend to be passionate individuals. They tend to find their own way with little direction and are frustrated that others can't and sometimes don't even want to perform at their level.

Author John Maxwell says it this way: "You can't make ducks soar with eagles and vice versa." And that's okay. We need both those who work on teams and those who need to soar alone.

Regardless of their frustration, the key here is early intervention to try to save them from themselves. At the first sign of this monster, don't hesitate to react. There is too much at stake.

Explain that top positions come with responsibility, including a responsibility to help develop others, and that there is a right way and a wrong way to do it. They're in a leadership role and need to start acting like it.

Their behavior will distract and tear apart the team. Eventually, left unchecked, the top producer could spend his or her time fighting with the rest of the team that has had it with the so-called leader's bad attitude.

In the end, you're left with an entirely dysfunctional team with members who spend most of their time fighting one another.

A word of caution: you are dealing with passionate people who sometimes don't make rational decisions. They may be prepared to walk over the whole mess. If they do, let them go. The entire team needs to know that no one, not even top performers, are above the team.

Take the case of wide receiver Brandon Marshall of the Denver Broncos, drafted in the fourth round of the 2006 draft as the 119th overall pick—hardly a pedigree.

Despite his less that heralded arrival into the league, Marshall quickly became Denver's top receiver and soon won one of the NFL's top honors with an invitation to Hawaii and a Pro-Bowl appearance.

As his stock began to rise, the young receiver demanded that his contract be reworked. When management refused, he demanded to be traded. That too was dismissed by Broncos leadership.

While in training camp, Marshall decided to make his dissatisfaction with all of this known. During warmups, with media cameras all around, while the rest of the team ran, he walked. He also batted down balls thrown his way and punted a ball away from a ball boy instead of handing it to him.

Denver's rookie coach Josh McDaniels had enough. He suspended Marshall for the two remaining preseason games for conduct detrimental to the team.

A month later, after winning a nationally televised game with some last-second heroics, his coach said Marshall had been nothing but a model citizen since his return.

Unfortunately, Super Monsters don't always die so easily. By the end of the season, Marshall would test his coach and teammates again.

During the final week of the season the team needed a win to have a chance to make the playoffs. Marshall came up lame during practice. It was unclear if the team's most talented player would be available for the most important game of the season.

Marshall told the media it would be a game-time decision and he would play if he could. He determined for himself that practice for the rest of the week was out of the question.

Evidently his coach felt something wasn't right, saying he would only field players he and the rest of the team could count on. Instead of waiting for game day, he announced midweek he was not only benching the star wide-out but deactivating him for the game, which guaranteed he would not play.

Marshall predictably protested his coach's decision in front of the cameras.

Denver lost the game and did not make the postseason tournament. The coach's decision will likely be debated for some time among fans. But one thing no one is debating, including Marshall, is who is running the Denver Broncos. By the way, Marshall is no longer with the team.

You'll recall that a manager I mentioned earlier was in the people development business.

If you are too, you'll quickly find your next superstar, and, at the same time, you'll have sent a message that no one, regardless of how good they are, will be allowed to threaten the well-being of the entire team.

Despite his monster slaying skills, Coach McDaniels was fired just as the 2010 season was winding down. His team totaled just three wins in twelve games that season, and monster slaying without results will get you benched regardless of your heroic efforts.

WORDS HAVE MEANING...
ACTIONS TOO

"Power is not revealed by striking hard or often, but by striking true."
—Honoré de Balzac

Is it true that words and actions have meaning? That might "depend on what your definition of is, is," as President Bill Clinton once put it. Fact is, what you do and what you say does have meaning and often both become food for the monsters.

I was once preparing to address the entire workforce of a client regarding a new initiative that was about to be launched. The plant manager said he would like a few minutes before the address to talk to the employees about a new incentive program.

The new plant manager explained the safety incentive program, which, unlike programs in the past, actually required employees to do something to achieve the incentive.

Before, under the old regime, not getting hurt was enough to qualify, which sent reporting of minor incidents underground for the purchase price of a coffee cup, ball cap, or jacket.

That's a problem because often minor injuries point to deficiencies that, left uncovered, could lead to major accidents or ones that end up taking someone's life.

If you pay people not to tell you the little stuff, they tend to clam up.

What's amazing is how cheaply they'll sell out. The resulting effect is often, "My personal safety for a coffee mug with the company logo." My goodness!

The new program (thankfully) was not based on simply not getting hurt. Workers had to actively do something to improve safety. They would have to, for example, lead a safety meeting, or identify an unsafe condition and a viable solution and so on.

For the effort, employees would collect points for each activity and then redeem them through a catalog for things such as power saws, high-end coolers, and other valuable merchandise.

Another way the new incentive program differed from those in the past was that only hourly people were allowed to participate. That was because of a change in the tax code. If supervisors (salaried personnel) took part in the program, the incentive would be considered compensation and therefore be taxable.

Don't you just love the Internal Revenue Service?

Management wisely limited the program to hourly employees, saving themselves an accounting nightmare and releasing their employees of an additional tax burden.

Think about it. "Thanks for participating in safety, now go to the catalog and pick out a skill saw or something else you'd like, and, by the way, in the next paycheck we'll be deducting an additional five to ten bucks from your paycheck for taxes."

A supervisor in the crowd spoke up. He wanted to know why he should get involved in safety if he could not partake in the new incentive program.

What was his motivation? How about the legal requirement that all supervisors take an active part in protecting their employees' health and safety? How about just good old "manning up" to take care of your people, supervisor? I was trying hard not to blow my lid; after all, this was not my fight.

About then, one of the guys sitting at the supervisor's table leaned forward, looked down the table at his supervisor and said, "Your incentive? How about this? Your incentive is I go home safe to my wife and kids. If that isn't enough incentive for you, we'll head to Walmart after work and I'll buy you a (expletive) cooler." I could have hugged that guy on the spot.

Later, in a rare move for me, I approached the plant manager with the suggestion that he terminate that supervisor.

He had undermined their health and safety process, not to mention the entire leadership group. In all my years of consulting I have never before or since seen a so-called leader make a statement with such disregard for his own team and the organization as a whole.

The plant manger was way ahead of me. The termination process was already underway.

Several months later I was back at that site doing some follow-up work and was stopped by an employee in the plant who wanted to know if I recalled the incident. I told him I would never forget it.

He went on to tell me that the supervisor had been fired shortly after the incident and was telling anyone who would listen that I was behind his termination.

That supervisor and I had actually had several disagreements of our own during our meeting. I frankly wish I could take credit for his termination. Fact is that train had left the station before I could release the break.

While I can't take credit for this one, the plant manager gets a gold medal for monster slaying.

Monster See, Monster Do
(Actions too, part 1)

Sometimes the monsters just sit back and watch our actions, looking for their next meal.

I was visiting a client in California and we had his counterparts in from all corners of the country for a meeting.

During the day, the site's general manager asked if it would be possible to give everyone in the meeting a chance to tour the fabrication shop.

His men had come up with a new approach to their business and he thought those in the meeting would be able to take it back to their shops. Sounded like a great use of time to me.

Before the next break he came into the room with a box full of safety glasses. As we went through the shop door, which had a sign reminding all of us that no entrance was allowed without personal protective equipment (eye and hearing protection), I was not the only one to notice that our impromptu tour guide was wearing nothing but his regular prescription glasses.

One of the men in our group asked if he was going to put on some real eye protection. The answer was, "No, my glasses will work just fine."

For the record, regular eyeglasses are typically not rated as safety glasses and even if his were, they did not have required side shields that would prevent something from reaching his eyes from an angle.

When this deficiency was pointed out, he said that we would not be near the grinder area except to pass through and, at that point, he would simply turn his body away from the work area. Nice try, I thought, except he's not accounting for ricochets.

Granted, the risk was statistically very low that he would lose an eye walking by just once. But his eye was not all that was at risk here.

Needless to say, he was not going to budge on this one. In other words, you've told the boss twice now that this is not acceptable behavior and twice he has said it's okay with a you-might-want-to-let-it-go glare.

As we entered the shop I was not surprised to see several employees with their safety glasses protecting their ball caps instead of their eyes. Of course when they saw us, everyone quickly removed the glasses from their caps and put them on.

There are times when you have to call it like you see it, and I knew once the words came out of my mouth it would likely cost me a pile of money.

Back in the meeting, with the general manager still in the room, I asked the group if everyone noticed how several people put their safety glasses on when we entered the shop. They had.

I asked them if we could just peek in the shop now without the people there knowing it, would they likely be wearing their glasses or not? Forget for a second they are required to. What would the behavior be?

The consensus was they would not be wearing them at all times as required.

I think everyone feared where this might be going, so I just said it for them. The reason why many were choosing to not follow the rule is that when the general manager walks into the shop and does not follow the rule, it sends the clear message that rules are not important.

If the rule were truly about eye protection, the manager himself would surely protect his own eyes. His actions left any logical person to conclude that it was not about eyes, it was about following the rules for rules' sake, or worse yet, only because the boss said so.

The latter two rationales are tremendously weak in terms of motivation. In the future, I went on, when someone has an eye injury in this shop, it will be determined it happened because the person was not following the rules. I would suggest instead that he was actually following poor leadership.

And that one example of not following the rules, I said, puts every rule into play in terms of whether it should be followed or not.

As you might have guessed, my little moment of truth was costly. I have not been invited back since.

Poor monster slaying can be expensive. Worse, I didn't solve the problem, did I? The manager was still setting a bad example.

That's Expensive Coffee
(Actions too, part 2)

Earlier, we talked about change at the top. All the moving around is a gold mine for good consultants. Over the years I have developed some strong relationships with some clients and have always been pleased when they invite me back to work again as they've moved on to new opportunities. It's the best compliment you can get in my business.

On one such trip I was visiting an old client at his new company. As I arrived, I entered the lobby to find a receptionist area that had all the markings of a company in cutback mode. The area had a sliding glass window where the receptionist used to sit and a note along with a phone directory. The instructions were to call your party and they would come and escort you in.

Paul, the plant manager I was meeting, was running behind and asked that I sit tight for five minutes and he would be up to get me.

As I was making myself comfortable, I noticed a coffee pot just beyond the entryway and decided I had plenty of time to enjoy a cup. As I was pouring my coffee, I noticed a sign clearly written by an angry person. It read something like this:

> This is NOT free coffee. This coffee is a community effort! If you are drinking it then you have contributed…if you have not, you should do so now!
>
> What you people fail to realize is that when we bought coffee around here, we were spending over 100 bucks a month!!!!

I was polishing off my second cup of self-determined free coffee when Paul met me in the lobby. Realizing he himself had only been on the job a short while, I asked him if he'd had a chance to read the angry coffee memo. Clearly he had not read it, because when he did he ripped it off the wall.

The next question was so obvious I couldn't stop myself. I asked in his estimation how much that memo had cost him.

His response was, "Hard to say. No telling how long it's been up and who has seen it or if there are others like it around the plant." He conservatively estimated it had cost him tens of thousands of dollars in production, good will, and morale.

Paul told me the message of the memo was loud and clear: his people were not worth a hundred dollars. He said, "If that's all it takes to provide coffee around here, I'll buy it myself."

It's hard to say how much that memo cost the company, but Paul was likely right that it cost a lot more than a hundred bucks. And it's not just the wording or tone of the memo, although that did make it a lot worse.

When you manage by spreadsheet—that is to say look to cut every cent you can—it sends a loud and clear message.

If the organization is having trouble, be upfront and frank about it. In the long run, you gain more respect for being straightforward, and the people who hold one of the major keys to turning things around will be more likely to pull in the direction you need.

By the way, when you start cutting things like coffee and office supplies, it might add up logically, but politically you can be dead wrong. In this situation, the coffee cut was just more feed for the monsters.

"Hey, how bad is it here? We can't even afford coffee?" Or, "Told you they don't care about us, we don't even get coffee now." Either way, the monster is fat and happy.

I'm not sure just how much good will Paul has been able to buy back at a hundred dollars a shot, but it's a priceless move in an effort to slay some monsters and inject some good old respect back into his organization.

Fact is, words have meaning whether written or spoken and people do listen.

We've all seen the memo around the break room announcing, "Your mother does not work here." First of all, I was not under any illusion that she did, and if you think my mother would clean up after me or you, well, you clearly don't know my mother.

How about this one? Once during a break I went into a client's men's room to be greeted by this sign: "Try aiming, we're getting a little tired of mopping up your pee!"

The actual thought of peeing on the floor (out of spite) crossed my mind. By the looks of things, several had chosen to act out on the same impulse I had been able to resist.

Seriously, men should not act like little boys, nor should they be talked down to as if they were.

Disrespect begets disrespect. And when we speak to adults as children, why are we surprised when some of them actually follow our lead and act like children?

By the way, another client approached the same problem with the same technique, different tact. Their sign read: "Gentlemen, please be mindful of the task at hand."

It's amazing how clean this men's room was in contrast to the other one. It was palatial. Yes, I just described a men's room as palatial.

I understand there are slobs among us. I understand there are people who steal others' lunches from the break room refrigerator. (Okay, I never understood that one; I'm just admitting it happens.) Those times call for some leadership, not a blanket belittling of the entire staff.

Communication across the organization, and especially from leadership, needs to be respectful regardless of subject matter, including "potty posters."

CASPER THE FRIENDLY MONSTER

"Casper the friendly ghost, he couldn't be bad or mean.
He'll romp and play, sing and dance all day,
the friendliest ghost you've seen."

Remember Casper? He never really had what it took to succeed as a ghost. Sure, he had the whole flying through walls thing down, but he just couldn't stop being everyone's buddy.

Don't get me wrong. I actually liked the little guy's contrarian nature, and I'm in no way suggesting leaders should be tyrants. But leaders are supposed to lead, not be part of the pack. It's fine to keep the friendship, just make sure everybody understands that things have changed at work and new standards are expected.

One of leadership's natural transitions may create its own monsters. Often promotions come from within and with them an opportunity for the development of a friendly monster: You're working on a team and you're promoted. You're no longer a teammate, you're now the boss. Many new leaders have a tough time cutting those collegial ties.

One of two paths is usually chosen. The first is sudden distance from the former teammates. One client put it this way: she decided to shut off all past friendships, hoping that her team would realize that her role had changed and that they would respect her and her new position.

That's not what happened. In fact, rather than respecting her, most judged her more harshly. They perceived that she had changed but not for the better. Most of them felt she had altered her personality along with her values and nearly all of them surmised that power had gone to her head.

The other end of the spectrum may be just as dangerous. An example is the friend of mine who swore his promotion would not change him or his friendships. The result was that some of his so-called friends began to take advantage of him, banking that their friendship would shield them from accountability.

It doesn't matter which of those two paths you choose. Both tend to end badly—either with a disgruntled, underperforming team that's so tight it operates poorly out of fear, or a team so loose it lacks direction.

When promoted from within, realize that things have changed. You are no longer a teammate. You're now the formal leader. Explain to your friends that you have to actually expect more from them to offset any perception that you take care of your own.

On and off the job relationships are different. Continue to enjoy the friendships you develop at work off the job, but while at work, make sure the new standard is understood and accepted. Talk it out first!

I've had the great joy of working for some wonderful people. I've become friends with many of them, and it's great to work for and with someone you truly enjoy spending time around. What concerns me is that someone always seems to take advantage of the friendly boss's good nature, especially when that boss is a former peer.

A sales team I once belonged to had a blast working together. We were young, making great money and having the time of our lives. Mark was our lead dog on the floor. He had an amazing gift for sales and leadership.

When we lost our manager, Mark was tapped as our new leader. It was well deserved and we were ecstatic.

Mark's talents for leadership were amazing, but as time went on, it became increasingly clear that some of his former peers, whom he now had to lead, were not exactly putting his success anywhere near the top of their own priority lists.

They started showing up late, leaving early, and taking two-hour lunches, extra mental holidays, and three-day weekends, all the while figuring their buddy would cover for them.

One even called Mark to bail him out of jail after a heavy night of drinking—while on company business—ended with a heated argument with police.

We all have some responsibility to protect our friendships and, in turn, our friends, but it's a two-way street. Anyone who uses friendship as a weapon (or in this case a shield) isn't much of a friend at all.

Mark soon found he was spending his time covering his so-called friends' transgressions rather than leading his team. These friends, of course, were not friends at all. Eventually they and Mark found themselves out of the company, and we lost a potentially great leader with a tremendous amount of talent, all because he allowed a few buddies to take advantage of their friendship with the boss.

On the occasions that I have worked for friends, I have made it a practice to never get too comfortable in that relationship while at work.

On the job, I tried never to forget who worked for whom and I tried never to take or even appear to take advantage of that relationship. By the way, off the job we remained friends and all that went with it.

I always wanted to remember who the boss was and, out of

friendship, never wanted to put them in the awkward position of having to defend me and or have it suggested they were doing so because of favoritism rather than merit. Quite frankly, to do otherwise would be unfriendly.

But not everyone respects the fact that the boss needs space, especially if he or she is your friend. So it's important that leaders maintain reasonable space to ensure no one misinterprets or abuses their friendly nature.

When leaders don't maintain that little bit of space, especially with those they used to work with, someone usually steps up and takes advantage of the situation.

If you really want to maintain a friendship once roles have changed, do away with the buddy system. Your real buddies can handle that, and if it's your buddy that's now the leader, go above and beyond the call of duty so she's not forced to defend your friendship or her judgment.

Now you can take this too far. One manager I worked with felt it was completely inappropriate for her to even appear to have anything that would even remotely resemble a friendship.

She insisted her people fear her and made sure they were unable to complete any emotional connection with her. She considered anything more unprofessional.

Her team had a difficult time buying into her vision because they were never allowed to buy into her. Eventually her team left her and she left the organization.

To lead you must inspire, to inspire you have to be liked, or at least respected. She was neither and thus failed to lead anyone to anything other than distrust and indifference to her. Not exactly a strong position to lead from.

AIM AT MONSTERS ONLY!

"Anyone can become angry—that is easy, but to be angry with the right person at the right time, and for the right purpose and in the right way—that is not within everyone's power."
—Aristotle

Years ago I worked for a gifted manager. She had a real talent for making work fun, challenging, and meaningful, but she also appeared to have a problem dealing with problem employees head on.

During our weekly staff meeting, there was generally some time reserved to correct a few team members' behavior. It was nothing major, mostly "a few of you are showing up late and leaving early" kind of stuff.

Problem was the corrective message was aimed at the entire team, despite the fact that only a few people were taking advantage of her generous management style.

When I pointed out privately that perhaps it would be more efficient if she talked directly to those bending the rules one-on-one, her response was, "I think they're getting the message."

I'm sure they were. The problem was, everyone, whether they

deserved it or not, had to endure the butt chewing. The violators were allowed to hide in the comfort of the entire group, and the unacceptable behavior continued despite the public chastisement.

Everyone in the group knew to whom she was referring and some became frustrated that those people were not being held personally accountable.

Take it one-on-one and then all doubt about who the guilty are will be removed without punishing the good along with the bad.

Those who are performing the way you desire do not need to hear you point out others' deficiencies. Trust me on this one. They're already well aware of the situation.

What they need to know is that all members of the team—especially those not even mustering the same effort—are being dealt with head on. Everyone needs to know that accountability exists.

Despite their bravado, monsters need a place to hide, and for a few, "the group" provides nice cover. Flush them out and take them on one-on-one. Your team will thank you for facing the monster head on.

THE MOTHER OF
ALL MONSTERS

*"Giving money and power to government is like giving whiskey and
car keys to teenage boys."*
—P.J. O'Rourke

I remember a business teacher I had in college—it's funny how you
never forget good ones—who spent most of his lectures talking
about the philosophical aspects of management as opposed to
management by spreadsheets and such.

He once asked the class if there was any such thing as a stupid
question, and we fell for it hook, line, and sinker. We responded
with the classic, "The only stupid question is the one that goes
unasked." He corrected the class with, "The world is full of stupid
questions. Your responsibility as a leader is to make the person
asking it not feel stupid." That, he said, is just one of the many
responsibilities you'll have as a manager.

His greatest lessons seemed to come not out of the text book
but from his own life experience.

He told us that as managers we would have a great deal of
responsibility. These responsibilities would include budgets,
production and quality demands, safety, environmental and ethical

responsibilities, just to hit a few.

He cautioned us that if we failed to manage these things well, the government would step in and do it for us, and he assured us we would not like the way the government would run our businesses. How true.

I've always thought the government is at its best when it comes to providing for the national defense, building highways, protecting us from bad people (foreign and domestic), providing for the common good, and not much else.

Come to think of it, you could hold a pretty good debate on whether the government has overachieved at anything lately that it's supposed to be good at.

I have met very few people, for example, who think a great way to improve their organization is by more government help. Quite frankly, most people in business would tell you the government has "helped" too much already.

And while, no doubt, government intervention is sometimes necessary and in some cases even desirable, a lot of its efforts have evolved into giant monsters terrorizing businesses.

If you want to know whose feeding this monster, just watch the news.

During the Enron fiasco I, like many, wondered what would happen to all those employees who lost their pensions after being duped by their managers' unethical and (we now know) in some cases illegal advice.

Employees were sold on the idea of investing in their own company, a company many of its executive group knew was about to financially implode. Price per share rose and some executives cashed out with healthy bonuses from stock options before the crash they knew was inevitable finally hit.

And it wasn't just Enron's employees who were hurt. Thousands of Enron's vendors were caught up in the mess created by Kenneth Lay and company. Many had much of their businesses invested in Enron and its service contracts.

This unethical behavior affects us all. Those in the market took a hit, investor confidence was rocked, the overall markets and economy were staggered, and the government monster had yet another ad campaign to take to the public about the "need" to enact new legislation that would supposedly get credibility back in financial reporting.

The Sarbanes-Oxley Act passed in the wake of the collapse of Enron, WorldCom, Tyco, and others now calls for fines, jail time, or both for executive leaders who knowingly sign off on bad numbers.

At the time of its passage, those who were for it argued it was necessary to restore confidence and accountability in the marketplace. Those against it claimed it would exacerbate an already overly complex regulatory environment and would dull America's competitive edge internationally.

I'm afraid examples of this type of unethical behavior are all around us, impacting us all. Take for example the sub-prime loan mess. Mortgage lenders were handing out high-risk loans to people who could not afford them, all under the direction of the Federal government and the Community Reinvestment Act (CRA) signed into law by President Jimmy Carter in 1977.

The Clinton administration in 1995 made various changes to the CRA that called for increased access and public underwriting of what we now call sub-prime mortgages.

I'm not dismissing the personal responsibility all of us have to read the fine print; however, one can't help but suspect it's in the fine print as part of an effort to be deceptive.

Millions of homeowners got caught up in the fiasco. Fannie Mae and Freddie Mac are in shambles and all of us taxpayers are picking up the bill on this one.

And you don't have to be politically astute to understand how things like what happened in the housing market and other misbehavior feeds the regulatory-control crowd in Washington.

A 700 billion dollar bailout (plus the 250 billion in pork Congress slapped on so they could "sell it" back home) resulted in capital being shut off to the people who need it the most to invest in the American Dream. Taxpayers were bailing out the mortgage business that was in turn shutting off access to loans to the very same taxpayers who were bailing them out.

It's a vicious game. The rich lenders get slapped on the wrists as their pockets are filled with cash and you and I pay the bill.

Fannie Mae, Freddie Mac, Lehman Brothers, AIG, Bear Sterns—all post Sarbanes-Oxley, so much for the government saving us from ourselves.

Interestingly enough, the last time interest-only loans were popular was just before the Great Depression. After the collapse of the housing market, along with the economy, new regulations were passed to hold lending institutions accountable.

Those practices were in place for more than fifty years with little problem until deregulation and the Fair Housing Act came along and converged, encouraging loans to previously unqualified borrowers so they could buy houses they could not possibly afford.

I'm not against deregulation. Less government control over our lives and decisions is in all likelihood better. You and I tend to make better decisions about how to run our lives than politicians. So when we fail through naivety or unethical behavior and we ask government to clean up the mess, we're creating a massive monster. By the way—NEWSFLASH! WE ARE THE GOVERNMENT!

Regulatory agencies like OSHA, the EPA, and many others are seen as cumbersome burdens by many, and in some cases they are. Just remember, all of those agencies can trace their birth to irresponsible leadership. As many in the safety business will tell you, all those laws and regulations (OSHA) are written in someone's blood.

TRILLIONS of our dollars are being handed out with reckless abandon and with little or, in some cases, no accountability. That unchecked largesse and the funding of Washington's massive appetite for pork is bankrupting our country and destroying our kids' future.

For those who think this is a right versus left issue, remember President George W. Bush and his administration oversaw the most expensive expansion of federal regulation in history.

While President Clinton cut the nation's regulatory staff by nearly a thousand folks, President Bush added over ninety thousand employees to the federal payroll.

When it comes to big government the monsters have no real party affiliation.

To state the obvious, there are some monsters we create that we will never be able to destroy. We might want to keep that in mind when we're deciding just how much to feed them. We all have a responsibility to conduct ourselves and our businesses with integrity. Each time we choose not to, even in the smallest way, we're contributing to the biggest monster man has ever created.

As a leader, keep your actions above board and expect nothing less from those who report to you.

Dishonestly is like a crack in the sidewalk. It starts out small, but once time, wind, rain, ice, and heat do their work, you're looking at a massive and expensive clean up like BP...oops, here we go again.

WHO MOVED MY CARROT?

"Men of integrity, by their very existence, rekindle the belief that as a people we can live above the level of moral squalor. We need that belief; a cynical community is a corrupt community."
—John W. Gardner

Some management teams I have encountered could be likened to a bad slot machine. Well, bad for the player (employee) at least.

They have a propensity to never pay off.

Like many of our other monsters, there are several ways you can recognize these beasts. The easiest tipoff is when they set a goal and hope no one hits it. I know it sounds strange—hitting a goal should mean growth for the company. But set a goal and then hope your team underachieves so you don't have to pay out any reward? What kind of goal is that?

I have seen actual pay plans that employees must sign as a condition of employment loaded with some pretty difficult goals but equally lofty rewards designed to incentivize the team to overachieve. I'm not talking about that. We often have to push to grow. What I'm talking about is when you get the push, hit the goal, and find out the rewards never materialize.

It goes something like this: "Well, we never intended or thought you could hit the goal, so we surely cannot pay it now."

In some cases, the incentive is simply not paid or arbitrarily lowered come payday, with no recourse on the employee's part.

It's hard to imagine why managers at an organization like this fail to understand why they are not successful and why they cannot get their employees on board with the overall direction of the organization.

In the few cases that it is too politically or legally hot to welch on the deal, the payoffs are loudly and begrudgingly paid and the pay plans are quickly adjusted to ensure the sales force will never again achieve that kind of reward. Problem is, the sales group is on a performance-driven pay plan. Create a pay plan that does not pay and sales people tend to quit selling, at least for you.

I've seen even worse. A friend of mine was a salesperson for an organization. Amy worked for a manager who constantly underperformed. In order to drive revenue and get some heat off of him, he devised the following scheme.

He would pay a bonus on every "preview" that went out the door to a client, regardless of whether it ultimately converted to a sale or not.

This organization sold workplace training primarily via videotape. A long-standing measure of productivity was how many of these previews each associate sent out per week.

The idea is, more previews equals more potential sales. It might have been sound thinking if said previews were qualified (meaning the potential customer was actually looking to buy something), but the vast majority of these were not.

Instead, potential clients were receiving unsolicited previews of products they might possibly use, but no one had any idea if they actually wanted them.

Shipping costs skyrocketed, inventory plummeted, and the salespeople were making money hand over fist off the new bonus plan without actually selling a thing. Sure, the salespeople should have shown some restraint, but asking salespeople to turn down free money is a little like asking me to keep an eye on your ice cream. It's a good bet there's going to be trouble.

With hundreds of unqualified previews out, the next pay period saw tens of thousands of dollars being, in essence, overpaid to the sales team. The jig was up.

The sales manager was fired, which he should have been. The bonuses, however, had already been paid, so the organization felt something had to be done about that.

The company decided it wanted its money back and began deducting a set amount from paychecks until the bonuses were paid back in full.

Many of the salespeople simply quit on the spot and pocketed the cash. So far so good, they ran off some unethical sales folks. The problem was, they were sent on their way with a nice little bonus.

The sales people who stayed were left holding the bag. Amy was a commissioned sales person. And if you've ever written your own paycheck in that regard, you know how difficult it can be during lean months.

Regardless of how good or poor the month was, Amy and others systematically had massive deductions taken out of their checks. At times they literally wondered how they would survive.

The remaining sales staff was punished for the acts of an incompetent manager. It seems to me the ultimate fault here was not even with the manager, but with an executive leadership group that put him in charge in the first place, then failed to remove him before it was too late.

It was also this executive "leadership" group that should have put someone in place to make sure the sales staff included the type of people the company would want around for the long haul, not the kind who would leave over a few thousand unmerited dollars.

In the end, Amy and others like her who chose to stay eventually paid back the company for its own lack of leadership.

Not surprisingly, the company that provided this example as well as the next one is no longer in the monster-creating business, or any other business for that matter.

If you refuse to kill monsters, don't be surprised if they end up killing your career or, in some cases, the entire organization.

A PLAN FOR MEDIOCRITY

"Unless you try to do something beyond what you have already mastered, you will never grow."
—Ronald E. Osborn

Early on in my career, I made a living in commissioned sales. It's not the easiest way to make your way, but I found working without a financial net was both rewarding and exciting. I've worked without a net to some degree ever since.

As one manager put it, "You're writing your own paycheck, so how much are you going to pay yourself?"

One of the most appealing aspects of this type of pay plan is that it's purely performance driven. You don't have to worry about carrying others, and what you get paid frankly depends on how much you're worth. Granted it's not for everyone, but I loved the challenge, and it was paying off.

One of the major challenges of this type of work is that the bar is constantly being raised. Quotas always go up, never down, but the upside is generally unlimited earning potential.

So when the new pay plans were rolled out about every six months, no one was surprised that the company expected more each time. While it came as no surprise, I was often taken aback by the number of people complaining that the new goals were unfair if not down right impossible to meet.

One of our senior sales reps had a different take, though. It never mattered what the pay plan was, how high the goals were, or what shape the economy was in; she always made it work and she always made money.

I decided early on that her attitude was the right one and tried to mimic it. Hoping they would not ask for more production was foolish. I decided that no matter what was rolled out, I would do as she did and focus on making my way to the top of the sales force and raking in the biggest bonuses.

This attitude adjustment on my part ushered in some exciting times. Along with my teammate, we began to produce revenue and generate compensation like I had never seen before. The sky was the limit, at least for the moment.

Now, keep in mind these adjustments in quotas were not exactly insignificant. They were generally in the range of 50 percent growth annually. As the quotas began to grow, the compensation did as well for those who could produce.

The problem ended up being a lack of a strategic growth plan combined with little marketing and no new products. At some point, sustaining 50 percent annual growth becomes impossible. In essence, management's strategy for growth was simply, "Salespeople sell more."

After a couple of years of this approach, we were in full growth saturation. With each pay plan calling for more growth than the last, fewer and fewer people were hitting their goals. It was quickly becoming a situation in which salespeople sold more than ever before and made substantially less money in the process.

Something else had developed during this period as well—the compensation cap. Regardless of how well employees performed, the amount of compensation was set in concrete. Hitting 100 percent of the plan earned an employee X dollars, but hitting 127 percent of the plan only earned him that same X dollars.

I had a plan to make more money for us all and went to my boss. My basic proposal was that if I doubled the new annual goal, which at that time was a million dollars, would she agree to pay me an amount above the fixed rate?

Realize this would be no small accomplishment. At the time, our top selling ticket was about $20,000. I was going to have to move a lot more inventory if I hoped to hit my number. Frankly, I wasn't sure I could do it, but if I could, what would it be worth to the company? "Nothing—we' were operating under a cap." I was told.

While I understood the cap philosophy, I openly questioned why management believed the sales force would be motivated to go above and beyond the call of duty if it didn't earn any extra compensation. My boss told me, "I would think you would do it for professional pride." Doubling one's revenue production over the course of a year takes a lot of effort, and pride alone is hardly compensation for the tremendous amount of work and time it would take to achieve such a lofty goal.

Keep in mind the original pay plan goal was already quite aggressive. My guess is that only about 20 percent of the team had any realistic chance of hitting the original goal.

I thought, "Surely they don't understand my proposal to double my own output." So I stated it as clearly as I could. I would hit 100 percent of goal or one million dollars in revenue. I would then be willing to go beyond the original goal and deliver another million dollars in revenue.

So I asked, "If I gave you a million dollars extra revenue, would you pay me a $30,000 bonus?" The answer was no.

As the years have gone by, the wisdom of that response and that plan remains lost on me. At a time when just about everyone was knocking on that same office door to complain that the company was being unreasonable with its original goal, here I was saying let's double that goal—just pay me a bit more for it.

Put yourself in the role of a manager. Would you rather have a team with people on it who say they can't, or would you rather have a team with people who say, "I can double that—what's it worth?"

By the way, my approach was not popular. Many took it as selfish, cocky, and proof that I somehow did not appreciate the company.

I never did achieve that self-suggested goal. Without compensation out there to chase, I quite frankly never saw the point in making the extra effort. I did hit 100 percent of plan and purposefully not much more.

Be careful about not rewarding those willing to go above and beyond. If you set the bar too low they just might clear it and nothing else.

I later heard that the company's chief financial officer complained that salespeople are only motivated by money. I'd have to agree, though I'm sure we'd view the same conclusion differently.

What he called greed, salespeople tend to call professional pride; pride that also adds tremendous value to the organization's bottom line.

Here's a business tip: if someone offers you a million dollars for thirty thousand, take it.

The point here is that if you have people who are willing to go the extra mile, reward their performance. If it's increasing sales, production, quality, or any other initiative, reward those who are willing to lay it on the line. Ultimately, it may be a very cost-effective way of sparking some innovation on your team.

A smart leader should reward employees who are eager to go the extra mile, not give them a reason to quit. At a time when employees are being asked to make concession after concession even as executives continue to give themselves multi-million dollar bonuses, something has to give.

I'd err on the side of those adding more value than those subtracting it in the form of large bonuses.

THE MONSTER OF
ALL MONSTERS

"He who fights with monsters might take care lest he thereby
become a monster."
—Friedrich Nietzsche

This identifying-monsters-in-others business is hardly rocket
science. It's easy for us to identify the shortcomings of other people,
and, no doubt, on our little journey together through these pages
you have, at times, conjured up an image of a co-worker or boss
you've encountered along the way.

While identifying and dealing with these monsters is critical
to achieving optimal organizational success, the most dangerous
monster of all may be the hardest monster to see. There's an Irish
proverb that says the longest journey a man will take is between his
head and his heart.

It's difficult to see our own shortcomings, but the truth is,
I know many of the monsters I've introduced you to on a very
personal level. With the help of others, I set out to kill my own
monsters—monsters that I had created and kept alive for years.
And no, I have not killed them all. I'm still on that head/heart
journey.

The first step was getting my heart to agree with my head that many areas in my life needed vast improvement. Frankly, some of my personal monsters were completely out of control.

I knew most of these monsters quite well. Other monsters who were hanging around unbeknownst to me were introduced to me by other Monster Slayers I encountered who cared enough to point out that I had lots of room to improve.

While it was tough to hear, I'm glad they had the courage and wisdom to coach me past my excuses. The monsters I protected then were inhibiting not just my professional growth but my personal growth as well.

I mentioned earlier that I had been fired several times and that eventually it became a positive experience for me. Initially though, I was hurt. I completely failed to comprehend everything that experience was ultimately going to teach me.

Working for one organization, I began to develop some of the passion I hope you found in these pages. Throughout my career, I believe my passion has been my ultimate strength. But there was a time it led to my downfall.

Passion can be frustrating when you want everyone to discover what you have discovered—some people just frankly aren't that interested.

Instead of accepting that others may never be as excited as I was about what we were doing, I set out to convert the entire organization to my way of thinking.

In short, I became a zealot. What a colossal mistake! It's great to have passion and conviction for what you do, but you still have to expect that not everyone will see things your way and you still have to respect their opinions.

I used every logical argument I could think of to make my case, but I found few takers. I then became argumentative and eventually dismissive; if others could not see the forest for the trees that was not my problem. In the end I became rude, unprofessional, and even combative.

I set out initially to show everyone the way. In the end, I simply sent them away.

After eight years in a career I loved, I was told that I was being let go. It hurt, and frankly it wasn't until years later that I accepted my fair share of the blame for getting myself fired.

It's always easier to point the finger at others, but it's not always the most productive approach. Looking in the mirror and holding yourself accountable for where and who you are, that's really what that head/heart journey is all about.

Look, there is a time and a place for righteous indignation, but there was nothing righteous about what I was doing. I let my frustration back then get the best of me. Instead of slaying that small monster, I fed it, and soon there were more monsters than I could control.

At the time, I blamed others for those monsters. Looking back on it now, their behavior may well have played a role in creating them. But I'm responsible for my monsters. I cannot control what others do with theirs. Worse yet, I could not even offer to help them with monsters they may have harbored because they were too busy dealing with mine.

I realized I had become the problem, not the solution. That was a difficult moment and, I believe now, a major turning point in my life.

I'm committed to being a lifetime Monster Slayer. I'm excited to share with others what I've learned. On a personal note, it's been interesting that for every monster I've taken on in my own life, I find another and another after that, sort of like layers of an onion. The more I peel the more I find.

These deeper monsters are stronger and more ingrained than those before them. They have been hiding and growing behind smaller monsters for a long time.

The monsters I battle today are more of a challenge than those of years past, and they should be if I'm making progress.

I know my personal battles as a Monster Slayer will never end, but as I look back to that morning when I was fired, I hardly recognize that man.

Today he's covered with battle scars. But he's stronger, kinder, humbler, and more passionate about his life and his work than ever. I'm better than I was and that is all I could ever hope for—to improve day by day.

Look, no one wants to hear they're not doing as well as they are capable of doing, including me, but the fact is, few of us are.

Stop right now and flip back to the table of contents. Remind yourself of the monsters we have uncovered. Nearly every one of them has lived within me at some point. How many reside in you? What message does that send to others around you? Not just at work, but in all aspects of your life, your home, and your community.

It would be a little disingenuous to assume the mantle of Monster Slayer when others see you keeping company with the beast. You'll be dismissed as I was. No one will take you seriously, and, trust me, that's a really lousy and lonely place to be.

Monster slaying, in the end, is about cutting away that which is stunting growth to help spur individual and, in turn, organizational growth. It's going to be hard to lead others to growth with an attitude that you've got it all figured out and have no room for improvement yourself.

When you stop growing, you lose the ability to help others grow. The monsters are fat and happy and we all suffer.

Just a reminder of a point I touched on earlier. Slaying monsters in yourself or others is a process. Don't expect to change the world overnight.

Don't be too hard on yourself or others when it doesn't just fall into place. I heard once, "The race is long and in the end it's only with yourself."

With effort and commitment you will become a Master Monster Slayer. You have what it takes or you never would have made it this far. Finishing this book should be telling you that you have what it takes.

Take your time. Be patient, dedicated, and purposeful, but also take pleasure in the fact that you are a noble Monster Slayer. Keep in mind that things of a negative nature in organizations tend to create their own gravitational force. Negativism is easily sustainable.

Things of positive nature like self-reflection and improvement and creating a positive culture around you take time to nurture and grow. It may take you awhile to convince others that monster slaying is worth the tremendous effort. Remember, the race is long.

Over time, you will help others find success and enrich their lives. Those around you will be better off because of your presence and you'll leave the organization better than you found it. Noble indeed!

ONE MORE TO KILL

"How soon 'not now' becomes 'never.'"
—Martin Luther

Writing this book has been a great joy. It has also been a lot of work, and it seemed, at times, as if I would never get it done. I spend two-thirds of the year on the road and am always busy. At times it even seems I'm too busy to do something I love, like writing this book.

Procrastination is a monster you must slay immediately in yourself. What's trapped in you right now waiting to get out? What dreams are you putting off until tomorrow?

Do me and yourself a favor. Put down this book right now and write down a list of things you've always wanted to accomplish. Prioritize them one through ten. Don't have ten? Then write down what you do have.

Give this some meaningful thought. If it takes a day or two, so be it. I'll be back here when you've completed your list.

Leadership expert John Maxwell, in his book *The 21 Irrefutable Laws of Leadership*, suggests you write down and rank the things that are required of you, give you the greatest return, and bring you the greatest reward. Then by applying the Pareto analysis, or the

80/20 rule, you focus on the top 20 percent in terms of importance and there you will find 80 percent of your return for your effort.

I read that one night after a friend, mentor, and fellow Monster Slayer told me that procrastination was a monster that had worked its way into the completion of this book. It's ironic that I read Dr. Maxwell's words shortly after my friend's encouraging nudge.

Writing this book has been just a small part of my life, but its return for the effort has been immeasurable. *Quit Feeding the Monsters* has been locked inside me far too long. There was always something else to do. Fact is, there always is. The question is, what should you be doing that you're putting off?

Now go back to that list. Take a look at the top 20 percent. Make a commitment to yourself to take a step each day toward that goal. Don't worry if the goal takes years to reach, just keep taking those steps.

Dr. Henry Cloud calls it "being the ant." Ants create tremendous communities in which thousands live, but they create it a micro step at a time and they never quit working toward their goal.

What's locked inside of you? In addition to becoming a Master Monster Slayer, what else do you want to do? What do you have to teach the rest of us? We could use your help, so get busy.

Our journey together is complete. Godspeed, noble Monster Slayer.

About the Author

J. Kevin Cobb has brought Advanced Safety Awareness Skills to some of the toughest environments imaginable across the United States, Canada, and Mexico.

For over a decade Mr. Cobb has studied not only the way of the monsters but their hunters as well.

When he's not out slaying monsters he's at home with his wife, Kristi, and their children in Texas. You can write J.Kevin at kevinc0930@sbcglobal.net or visit www.quitfeedingthemonsters.com